Mushroom

A Nutritive Food and its Cultivation

The Authors

Prof. Suresh Borkar was borne on 7th July, 1956 in Bhandara district in Maharashtra state, India. Obtained B. Sc. (Agril) degree from Dr. Punjab rao Deshmukh Krishi Vidyapeeth, Akola, Maharashtra in 1977; M.Sc. from IARI, New Delhi in 1979; Ph. D. from IARI, New Delhi in 1983; Post Doctorate from INRA, France in 1984 and D. Sc. from University of Washington, U. S. A. in 1999 in the subject of Plant Pathology.

After returning from France, joined as Assistant Professor of Plant Pathology in 1985 at Jawaharlal Nehru Agriculture University, Jabalpur, M.P. Selected to the position of Associate Professor of Plant Pathology by Mahatma Phule Agriculture University, Rahuri in 1989 and joined this position, become Professor in 1994 and Head of Department of Plant Pathology and Agriculture Microbiology in 2005 in the same university. Also served as Dean, College of Agriculture, Kolhapur and Dean, Post Graduate Institute, Mahatma Phule Agriculture University, Rahuri during 2013. Presently Heading the Department of Plant Pathology at MPKV, Rahuri.

Guided more than 25 post graduate students for their masters and doctoral degree in Plant Pathology and Agriculture Microbiology. Recipients of 18 National and International Awards and accolades. Published more than 77 research paper in Indian and foreign journals abroad in the subject of Plant Pathology and Agricultural Microbiology. He has one patent to his credit. Published 3 university publication and one book (by Pointer Publishers, Jaipur). Chaired many scientific sessions in several workshops/seminars and is on selection committees of different Agriculture Universities. Visited many Universities/Institutions in Europe. He is Fellow of Indian Phytopathological Society and a Fellow of Eurasian Academy of Environmental Sciences.

Dr. Nisha Patil was born on 10th Sept, 1985 at Nagpur. She has completed her graduation from Dr. Punjabrao Deshmukh Krishi Vidyapeeth, College of Agriculture, Nagpur in 2008 and post graduation from University of Agriculture Sciences, Dharwad, Karnataka in the subject Agriculture Microbiology in 2010 and Ph. D. from Mahatma Phule Krishi Vidyapeeth, Rahuri in 2013 under the supervision of Prof. S.G. Borkar. She work on the use of rooting inducing microbes in Agriculture and identified the rooting inducing microbes which can replace the use of plant growth hormones in MS media. Her areas of interest is in microbiology, tissue culture, molecular genetics and biotechnology.

Mushroom

A Nutritive Food and its Cultivation

Authors

S.G. Borkar

Nisha Patil

Department of Plant Pathology and Agril. Microbiology
Mahatma Phule Agriculture University, Rahuri
District Ahmednagar, Maharashtra (India)

2016

Daya Publishing House®

A Division of

Astral International Pvt. Ltd.

New Delhi – 110 002

Cataloging in Publication Data--DK
Courtesy: D.K. Agencies (P) Ltd. <docinfo@dkagencies.com>

Borkar, S. G. (Suresh Govindrao), 1956- author.
Mushroom : a nutritive food and its cultivation / authors, S.G. Borkar, Nisha Patil.
     pages cm
     Includes bibliographical references and index.
     ISBN 978-93-5130-949-9 (International Edition)

     1. Mushrooms. 2. Mushrooms--Nutrition. 3. Mushrooms--Technological innovations. 4. Mushroom industry. 5. Mushroom culture.   I. Patil, Nisha, 1985- author. II. Title.

QK617.B67 2016   DDC 579.6   23

Published by   : **Daya Publishing House®**
                A Division of
                Astral International Pvt. Ltd.
                – ISO 9001:2008 Certified Company –
                4760-61/23, Ansari Road, Darya Ganj
                New Delhi-110 002
                Ph. 011-43549197, 23278134
                E-mail: info@astralint.com
                Website: www.astralint.com

Laser Typesetting   : Classic Computer Services, **Delhi - 110 035**

Printed at   : Replika Press Pvt. Ltd.

Acknowledgement

The authors acknowledge the scientific contribution of all those whose work and references are quoted in this book. Similarly some of the photographs of mushrooms and mushroom recipes are also taken from internet and are thankfully acknowledged for the contributors to internet. Wherever the contributors are known on net, their names are mentioned at respective place, nevertheless we acknowledge all those whose net photograph and text are included in this book.

The authors thank Smt. Kavita Sandip Dhawade for her painstaking effort in the typing and manuscript setting work.

Prof. S.G. Borkar

भारत सरकार
कृषि अनुसंधान और शिक्षा विभाग एवं
भारतीय कृषि अनुसंधान परिषद
कृषि मंत्रालय, कृषि भवन, नई दिल्ली 110 001

GOVERNMENT OF INDIA
DEPARTMENT OF AGRICULTURAL RESEARCH & EDUCATION
AND
INDIAN COUNCIL OF AGRICULTURAL RESEARCH
MINISTRY OF AGRICULTURE, KRISHI BHAVAN, NEW DELHI 110 001
Tel.: 23382629; 23386711 Fax: 91-11-23384773
E-mail: dg.icar@nic.in

डा. एस. अय्यप्पन
सचिव एवं महानिदेशक
Dr. S. AYYAPPAN
SECRETARY & DIRECTOR GENERAL

Foreword

The nutrient-rich mushrooms are presently known for its medicinal values. This has led to a common saying that "mushroom in a week keeps cholesterol and blood pressure away". However, mushrooms are not available like vegetables in a market throughout the year, as they are cultivated on a small scale, mostly by individual households leaving an opportunity for integration of mushroom cultivation technology for commercial production vis-à-vis socio-economic development of mushroom growers.

The book authored by Prof. S.G. Borkar and Dr. Nisha Patil on "Mushroom: A Nutritive Food and its Cultivation", summarizes the historical background and development in mushrooms science, its nutritive and medicinal values, edible and poisonous mushrooms, mushroom cultivation technologies for several edible mushrooms, its processing and storages. Several mushrooms recipes are also given for homemakers to keep their families healthy.

I believe, it is a step forward to disseminate the scientific knowledge on mushroom and its cultivation and a good and complete source of information and knowledge of mushrooms to students, entrepreneurs, researchers and common homemakers.

(S. Ayyappan)

Preface

Mushroom, a nutritious vegetarian has a great potential as on farm and off farm industry.

The Global mushroom production in 2009 was 2.4 million tonnes which is growing at the rate of 7 percent. China with 1.7 million tonnes production accounts for nearly 70 per cent of world production. Other major mushroom producing countries are USA, Netherland, Poland, France, Italy, Indonesia and Germany. India with 1.5 per cent contribution ranks eighth in the global mushroom production.

The mushroom consumption is mainly concentrated in six countries particularly USA, Germany, UK, France, Italy and Canada, consuming 85 per cent of world consumption. The varieties of mushroom cultivated internationally are button (31 per cent), shiitake (24 per cent), oyster (14 per cent), black ear mushroom (9 per cent), paddy straw mushroom (8 per cent) and milky/others.

India's production of mushroom was 1,13,315 tonnes in the year 2010. Punjab, Uttarakhand, Haryana, Himachal Pradesh, Tamil-Nadu, Orissa, Andhra, Delhi and Maharashtra are the major mushroom producing states. Punjab, Haryana and Himachal Pradesh accounts for nearly 90 per cent of country's production while Punjab alone contributes to 50 per cent of the total production. Total mushroom export from India in 2009-10 was around 11,000 tonnes valued at Rs. 66 crore. Major export destinations for Indian mushroom are USA, Israel and Mexico. Button mushroom account for approximately 95 percent of total mushroom export. Mushroom export from India is in two forms- fresh and processed and is subjected to no- tariff barriers and thus the export is fluctuating. Although the current share of India in world export is less than 1 per cent, India has a great export potential.

India's per capita consumption (20.25 g) is comparatively low as compared to Europe and U. S. A. (2 to 3 kg). The domestic demand is growing at a rate of 25 per cent.

Mushroom is efficient means for conversion of agricultural wastes into valuable protein and present huge potential for generating additional income and employment. The domestic marketing channels lack adequate price support faced with erratic demand and supply. Lack of trained manpower is among other constraints in mushroom industry. Therefore, detailed scientific knowledge of mushroom and its cultivation is a pre-requisite to start this entrepreneurship. The present book, mushroom a nutritive food and its cultivation is a step forward in this direction.

Prof. S.G. Borkar

Nisha Patil

Contents

1

Introduction to Mushroom

Mushrooms are the edible fruit/fruiting bodies of the edible fungi having high nutritive and pharmaceutical values with delicacy. In common-mans word, as brinjal, tomatoes and cucurbit fruits are the fruits of respective vegetable crops, mushrooms are also the fruits of edible fungi.

The mushrooms have been considered a delicacy for thousands of years. We find reference of mushroom in the ancient Roman and Greek literature. A historical record about the beginning of mushroom culture is insufficient and the indications of prehistoric people using mushroom fungi are very rare. In Central America and Gautemala, mushroom –shaped stone carving have also been found. These belong to Mayan period and these were used in ceremonies. Mushroom has appealed to different people in different ways. They are object of beauty for artist and for medical people they are source of drugs. The architects have constructed minarets, temples and columns in its shape. Jewellers have made expensive pieces on mushroom design. Designers have reproduced mushroom on fabrics. Bulgaria has issued a series of nine stamps on mushrooms.

Among the hundreds of thousands of fungal species and the tens of thousands of those recognizable as gilled mushrooms, only about 100 would qualify as poisonous of which only about 10 are deadly. Nonetheless, poisonous mushrooms cast pallor on the consumption of fungi as food; the epithet toadstool elicits emotions that inappropriately and unfortunately associate the innocent mushroom with loathsome warty uncleanliness and therefore promotes mycophobia, the fear of fungi. The word toadstool itself has an etymology that is suggestive of Stygian origins; it is a calque word of German provenance. "Todesstuhl" means "death stool." One reason for mycophobia is identification since many of the deadly variants are similar in appearance to those that are edible; in many cases only an expert can tell them apart and even they can make mistakes. Another reason is indication; there are no sine qua non proofs of mushroom toxicity, poisonous mushrooms do not all blacken silver spoons and do not all taste bitter, popular nostrums aside.

History has impugned the mushroom as the source of the poison that has dispatched any number of notables, among them Claudius, the fourth Roman Emperor. The perpetrator is alleged to have been his fourth wife Agrippina who wanted her son Nero to succeed to the throne in lieu of Britannicus, the son of Claudius by his third wife Messalina (who had incidentally been put to death for her part in a previous assassination attempt). The death is recounted by the philosopher Seneca the Younger, only two months after the event occurred. According to his account, it happened quite quickly, the onset of illness and death being separated only by about an hour. The idea that mushrooms were the cause is perpetrated in later writings by Tacitus, Suetonius and Dio Cassius, the latter adding a description of Agrippina serving Claudius a plate of mushrooms. The mushroom death of Claudius is almost certainly apocryphal, as deadly mushrooms are relatively slow to act; those that act rapidly generally cause gastrointestinal distress that is rarely if ever fatal.

It is more likely the case that a mushroom poisoning occurs as a result of volition, the decision to consume a mushroom of dubious identification resting with the victim. There have been occasions where arrogant decisiveness has overridden caution. The most notable case is that of Johann Schobert, a composer who was employed by the Prince of Conti in Paris in the latter half of the 17th Century. He wrote harpsichord concertos, opera and sonatas that purportedly served as the basis for some of the later work by Mozart, his contemporary. Schobert may have had a talent equal to that of Mozart; we shall never know, as he succumbed to mushroom poisoning. According to the historical account, he had gathered some mushrooms in Pré-Saint-Gervais near Paris with his family and proceeded to a restaurant to have the chef prepare them. When he was told that they were poisonous, he proceeded to a second restaurant with like result. Undeterred, he went home to Paris and made mushroom soup for dinner. He was joined in death by his wife, one of his children, and a friend, a doctor; fittingly, it was the doctor who had proffered the mushroom identification in the first place.

Mushroom poisoning is generally categorized into four types according to the symptoms that result. It is not practical to use the type of toxin as a basis for classification, since there is a paucity of knowledge about the nature and chemistry of fungal toxins in general; this is particularly true for fungi whose consumption may yield an unpleasant though not fatal result. In addition, for all practical purposes, the identification of the mushroom that caused the condition under evaluation is usually a matter of conjecture since the victim has succumbed to the condition after having eaten the evidence. The four types are protoplasmic poisons, neurotoxins, gastrointestinal irritants, and those that are toxic only in combination with other substances, notably alcohol.

The earliest word in Sanskrit for mushroom is *Ksumpa* and in present day Hindi, it evolved to *Kumbhi*. The invasion/migration of the Aryans into the Indian subcontinent, which took place around 1500 BC, might have carried with them an intoxicating drink Soma and this was mostly used in Aryan religious rites. Soma in the *Rig Veda* refers to a mushroom (*Amanita muscaria*). The earliest cultivated edible fungus was a species of *Auricularia*, which was cultivated about 1000 years ago in China. The next species to be cultivated was *Lentinus edodes*, almost 900 years ago also in China.

The earliest description and Know-how of growing mushrooms was written by Toune Forte, a Frenchman, and it was published in Paris in 1707. The method described by him is remarkably similar to that employed today. In fact, no radical change took place until 50 years ago. In 1800, the French started growing mushrooms underground in the querries, around Paris on horse manure, which was staked in heaps and allowed to heat up naturally. The resulting compost was laid down in long ridges and inoculated with spawn dug from meadows. Refined methods for mushroom growing were introduced as a result of the investigations of the French mycologists.

By the end of 19th century, French mycologists, among them Matrochot and Costanin (1894) succeeded in solving two important problems in mushroom cultivation. Firstly, they discovered the cause of the mushroom disease and started fumigating with sulphur and secondly they were able to germinate spores for the purpose of obtaining spawn. This was probably the most important contribution to scientifically controlled mushroom growing. Duggar an American in 1905 perfected a method of making pure culture spawn from mushroom tissue. This method was exploited immediately, as now it was possible to select and guarantee a particular strain spawn. From this onward, mushroom growing began to develop into the highly scientific industry that it is today.

In India, cultivation of edible mushroom is of recent times, though methods of cultivation for the same have been known for many years. In 1886, some specimen of mushroom were grown by N. W. Newton and exhibited at the annual show of Horticultural Society of India.

In 1908, a thorough search was initiated by David Prain for edible mushrooms as it attracted his attention because it was used as a food by the poor in the famine striken areas. Sinden (1937) observed that about one third of monospore cultures of *A. bisporus* were incapable of producing fruiting bodies. In 1939 experimental cultivation of paddy straw mushroom was done by DOA, Madras. Padwick (1941) tried cultivation of *Agaricus bisporus* all over the world with great success but it often remained unsuccessful in India and pointed out the possibilities and obstacles in its cultivation, whereas, Lambert observed the temperature between 50-60°C to be suitable for the preparation of good quality of compost. Thomas *et al.* (1943) gave full directions about the cultivation of paddy straw mushroom in Madras. In 1947 Asthana got better yield of paddy straw mushroom by adding powered red gram dal to the beds. He found April to June as the best period for the cultivation of the Paddy straw mushroom. Sinden and Hauser (1950-1953) developed short method of compost preparation. During 1961 first serious attempt on cultivation of *Agaricus bisporus* was started at Solan, under a joint scheme of H.P. Government and ICAR. Bano and

Srivastava (1962) reported mass production of mushroom on straw based substrate. They reported increased yield of *Pleurotus* species when paddy straw was used as substrate. Mantel (1965) guided construction of growing facilities and research on spawn production, synthetic compost etc. at Solan (HP), India. ICAR coordinated scheme on mushroom was started in 1971. W.A.hayes (1969), advised on improved methods of compost preparation, pasteurization, casing, environmental management for the yield enhancement etc. while working at Solan (HP), India. In 1974 cultivation technology of *Pleurotus* sajor-caju on banana pseudostem was perfected for the first time by Jandiak at Solan (HP), India and now this species is popular throughout the world. In 1982, ICAR sanctioned the creation of National Centre for Mushroom Research and Training (NCMRT) during VIth Plan with the objective of research and training and now its name is NRCM. During 1997, Solan was declared as Mushroom city of India.

Commercial mushroom growing was first initiated in Chail near Shimla by the name Tegg's mushroom farm in eighties; and later on it spread to J and K. Uttaranchal, U.P., Punjab, Haryana, Tamil Nadu. Maharashtra etc. In Haryana, commercial mushroom cultivation was started in early eighties in the Bhadana village of Sonepat District, and now the Haryana State is a leading producer of white button mushroom under seasonal conditions using low cost technology.

2
Nutritional and Medicinal Values of Mushrooms

Mushrooms with their flavour, texture, nutritional value, very high productivity per unit area and time, less dependence on land and ability to grow on a variety of residual agricultural wastes, have rightly been identified as a food source to fight malnutrition in developing countries. It has been estimated that the annual production of crop residues in India is more than 300 million tones. Though, the statistics about the ways in which these crop residues are being utilized are not available, but a sizeable portion is being either burnt or left to rot. Burning of crop residues, besides posing environmental problems, is a waste of energy and poor utilization of crop residues. These crop residues could be converted into food, feed and fuel by mushroom cultivation.

Mushrooms have been recognized by FAO as food contributing to the protein nutrition of the countries depending largely on cereals. Mushrooms like other vegetables, contain about 90 per cent moisture and are basically low calorie food (about 30 calories/100gm. fresh weight). Total carbohydrate content is 4-5 per cent, which consist of chitin, hemicellulose and glycogen. Very little amount of free sugars (0.5 per cent) is present. Starch is absent. The fat content is very low (0.3 per cent), but it is rich in the linoleic acid (70 per cent). Cholesterol, the dreaded sterol for heart patient, is absent and in its place ergosterol is present, which can be converted into vitamin D by human body. Fibre content is high, which is helpful in excretion of waste and prevention of constipation.

Vitamins A, D, E, K are almost present in mushrooms. These are rich in the vitamin of B-complex group and vitamin C, which are present in significant quantities, which include thiamine, riboflavin, niacin, biotin and pantothenic acid. Folic acid and vitamin B-12, which are normally absent in vegetable foods, are present in mushrooms. They are also fairly good source of vitamin C (4-8 mg/100g). Phosphorus and potassium are present in high quantities. They are comparatively deficient in calcium and iron, but the latter is present in available form. High K: Na ratio present in mushrooms is desirable for the patients with hypertension.

Table 1: Nutritional Values of Edible Mushroom vis-a-vis Common Vegetables (g/100g fresh wt.)

Mushroom/Vegetable	Moisture	Protein	Fat	Carbohydrate	Fibre	Ash	Calories
Agaricus bisporus	90.1	2.9	0.3	5.0	0.9	0.8	36
Volvariella volvacea	90.1	2.1	1.0	4.7	1.1	1.0	36
Pleurotus sajor caju	90.2	2.5	0.2	5.2	1.3	0.6	35
Cabbage	91.9	1.8	0.1	4.6	1.0	0.6	27
Cauliflower	90.8	12.6	0.4	4.0	1.2	1.0	30
Potato	74.7	1.6	0.1	22.6	0.4	0.6	97

Mushrooms have long been considered to have medicinal values. In fact, the early herbalists were more interested in medicinal properties in mushroom than in their basic value as a source of food. In 1991, according to an estimate the medicinal products of mushrooms generated 1.2 billion dollars. In Japan, the annual sales of a top selling cancer drug, 'PSK' from mushroom (*Cariolus versicolour*) were totaling 358 million dollars and accounted for 25 per cent of the country's total sales of anti cancer drugs. Chang and Buswell have coined the term 'mushroom nutraceuticals' for the medicinal preparations from mushrooms.

Due to unique chemical composition, mushrooms are suited to specific groups suffering with some ailments/disorders. As a low calorie high protein diet with almost no starch and sugars, mushrooms are the "delight of the diabetic'. Due to high K:Na ratio, few calories and low fat (rich in linoleic acid and lacking cholesterol), mushrooms are suited to persons suffering from obesity, hypertension and arthrosclerosis. Alkaline ash and high fibre contents make them suitable for those suffering from hyper acidity and constipation. Besides this, mushrooms are rich in vitamins; folic acid and available iron that reduce the degree of pernicious anaemia. Non- soluble polysaccharides of the several mushroom species have been shown to display anti-tumor activity. Some of the mushroom species display anti-bacterial and anti-viral activities also.

Some mushrooms produce toxins, which destroy nematodes. It also serves as therapies for hyperlipemia and diabetes and are also selective inhibitor of myxovirus multiplication. Thus, mushroom bears a number of useful functions for human health, beside their contribution to human nutrition. Though, few commercial mushroom products for medicinal purpose are available, but a day is not far to claim further increase in market value of mushrooms.

Table 2: Medicinal Value of Mushroom

Composition	Suitability
☆ Low calorie and high protein diet	☆ Diabetic patients
☆ Almost no starch and sugar	
☆ High K:Na ratio	☆ Obesity
☆ Low calorie and low fat (lacking cholesterol)	☆ Hypertension
	☆ Arthrosclerosis
☆ Alkaline ash and high fiber content	☆ Hyperacidity, Constipation
☆ Rich in Folic acid and available Iron	☆ Anemia
☆ Anti viral properties	☆ Recent reports of effective against AIDS
☆ Anti bacterial properties	virus
☆ Anti tumor properties	
☆ Rectification of renal properties	☆ Extend life span of chronic renal failure patients

Table 3: Medicinal Values in different Edible Mushrooms

Sl.No.	Mushroom Species	Medicinal Values
1.	Pleurotus ostreatus	Anti-tumour
2.	Pleurotus griseus	Antibiotic
3.	Pleurotus japonicas	Anti-leukemia
4.	Pleurotus sajorcaju	Cardiovascular and Renal effect
5.	Agaricus bisporus	Anticancer and antitumor effect
6.	Volvariella volvacea	Antifungal, antibacterial, antiprotozoal and anti- atherosclerosis.
4.	Lentinus edodes	Inhibits growth of subcutaneous sarcoma. Decrease hypertension and Plasma cholesterol level.
5.	Auricularia auricular	Anti ulcer effect
6.	Flammulina velutipes	Anti hypertensive and reduce plasma cholesterol
7.	Coriolus versicolor	Anti-tumours
8.	Schizophyllum commane	Effective against tumours and carcinoma

Medicinal Products of Mushrooms in the Market

Concord Sunchih

Concord International Trading Pvt. Ltd., Australia. Prepare this from fresh cultivated *Gandoderm* fruit bodies. The drug improves immunity,antihypertension, hepatitis, bronchitis, piles, asthma and diabetes.

Dosage

1 to 2 caps twice daily for health and 3 caps thrice daily for illness.

Griferon

Maitake products, Inc. New Jersey, USA, prepare this from *Grifola frondosa*. It combines well with conventional chemotherapeutic drugs against cancer.

Dosage

5-6 drops TDS for general health and for therapeutic dose 0.5-1 mg/kg body weight TDS.

3

Morphology and Classification

Mushroom is the fruiting bodies of the fleshy fungi and belongs to different classes of fungi. Mushrooms consist of two portions –one the fruiting bodies or the mushrooms itself and the other is thread like structures growing extensively in substrate comparable to the roots of higher plants. These thread like structures are known as mycelium which supplies nutrients from substrate to fruiting bodies. Most of the mushroom have caps and stalks whereas, there are some other varieties with different shapes and sizes and are devoid of stalks.

3.1. Parts of Mushroom

The parts of mushroom are depicted in Figure 1.

1) Cap or Pileus

It is the expanded portion of the carpophores which may be thick, fleshy, membranous or corky and varies greatly in shape, size and colour. The surface of the pileus may be smooth, hairy or rough. This is the most visible part.

2) Gills or Lamellae

They are situated on the underside of the pileus starting from the apex of the stalk and radiate out towards the margin. These gills bear spores on their surface and exhibit a change in colour corresponding to that of the spores.

For instance, in *Agaricus bisporus* the colour of the young carpophore's gill is pink. With age the colour changes to dark purple, brown or nearly black due to the changing colour of the spores. The attachment of the gill to the stipe helps in the identification of the mushroom. When the gills do not touch the stipe or only to do so

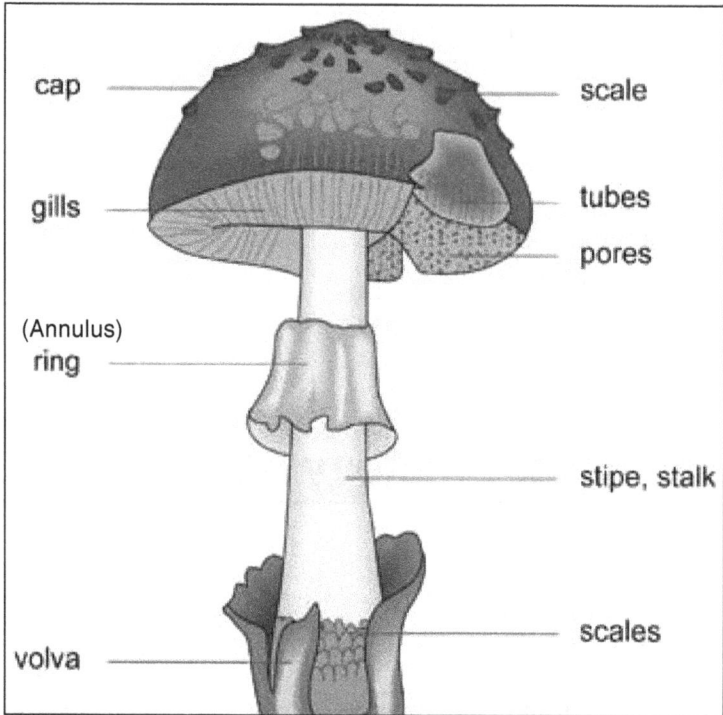

Cap

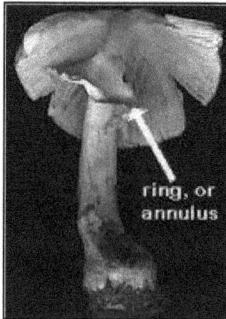
Annulus

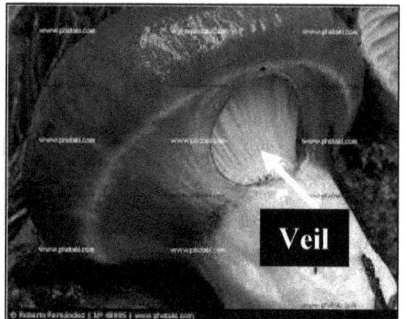
Veil

Gills

Stipe

Volva

Figure 1: Parts of Mushroom.

by a fine line they are known as free. When they are attached directly to the stem, forming nearly a right angle with the latter, they are termed as Adnate. If the attachment is only by a part of the width of the gills, they are Adnexed. When the gills extend down the stem to a greater or lesser degree they are known as Decurrent and when they are near the stalk in a deep notch they are termed as Sinuate (Figure 2).

Structure of the Gills

The centre of the gill is made up of mycelia thread known as trama. These threads may run parallel to each other or may be interwoven. Accordingly, the cells may be long or short. Towards the outside of the trame, the cells branch into short cells forming a thin layer, the sub-hymenium. The sub-hymenium give rise to long club-shaped cells which are parallel to each other and a right angle to the surface of the gills. These club-shaped cells are called basidia which have 2 to 4 spine – like projections called sterigmata, on which basidiospores are borne. Some inflated bladders like projecting beyond the basidia are the cystidia and those resembling basidia are the paraphyses. The cystidia, paraphyses and the basidia together constitute the hymenium. The basidiospores can be of different colour and shape with rough or smooth margin. The spores are very small and are not visible with the naked eye, but in mass they give a dusty appearance. The spores are the productive bodies and are produced in large numbers by a single carpophores. These spores either fall on the parent host or on the ground or are dispersed by wind currents. When spores fall on a suitable medium they germinate under favourable conditions.

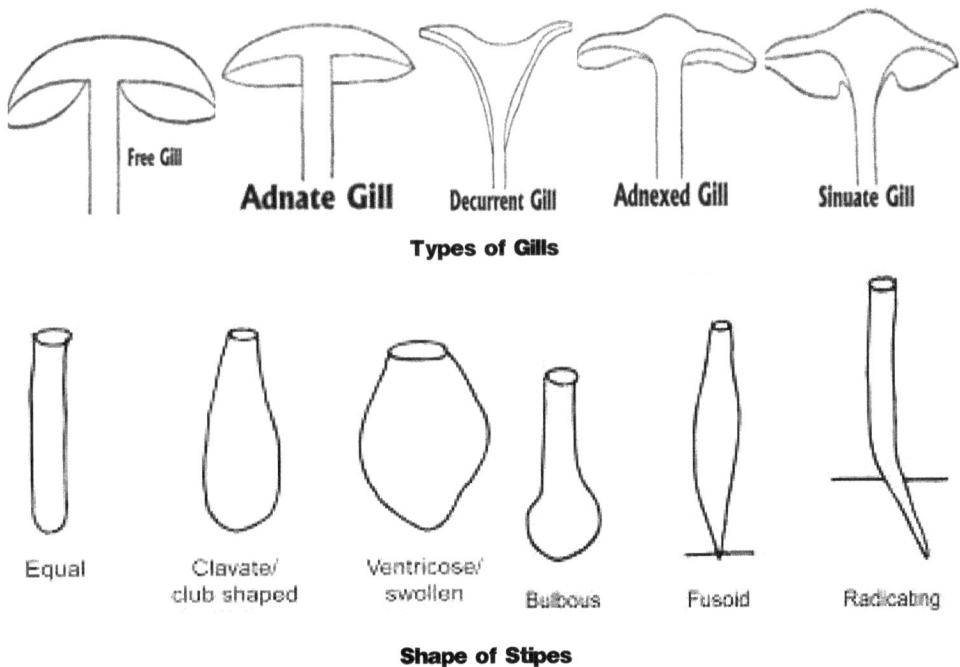

Free Gill   **Adnate Gill**   Decurrent Gill   **Adnexed Gill**   **Sinuate Gill**

Types of Gills

Equal   Clavate/ club shaped   Ventricose/ swollen   Bulbous   Fusoid   Radicating

Shape of Stipes

Figure 2: Types of Gills and Stipes.

3) Veil

In young fruiting bodies the gill remains covered by a membrane that extends from the margin of the cap (pileus) to the stipe, and is called veil. As the cap extends membrane breaks, some portion remaining attached to the margin of the cap while the other may form a ring on the stipe which is termed as the annulus. It is very delicate and can easily be rubbed off or may even be washed away by rains.

4) Annulus

A ring formed around the stipe is known as annulus. Annulus may be present or absent in different varieties.

5) Stipe and Stalk

The stalk supporting the pileus is also known as stipe or stem. Its presence or absence and mode of its attachment to the cap are an important character for identification of the mushroom. Mostly the stem is centrally attached to the cap, but in some cases the attachment may not be exactly in the middle but lateral, then it is known as Eccentric. The stipe may be solid, fleshy, or hollow. In shape the stem can be cylindrical, *i.e.*, having the same diameter throughout or spindle shaped, *i.e.* swollen in the middle and tapering towards both the extremities, or club shaped when it enlarges towards the top and tapers towards the base into a root-like form. The bulbous stipe is that in which the base shows a sudden enlargement whereas in marginate the stipe base widens into a sort of saucer with well defined margins (Figure 2).

6) Volva

Initially the entire fruit body prior to differentiation is covered by a universal veil. As the carpophore extends this veil breaks and may remains as a cap like structure surrounding the base of the stipe, or known as volva.

On the basis of the presence or absence of annulus and volva (Figure 3) the mushrooms can be put in the following four categories:(1) both annulus and volva present *e.g. Amanita*, (2) only annulus is present and volva is absent, *e.g., Agaricus*, (3) in which only volva is present and annulus is absent *e.g. Volvariella*, and (4) both annulus and volva are absent, *e.g., Marasmius oreades*.

3.2. Classification

The mushrooms have been classified by different workers both for identification and taxonomy purposes.

A) Occurrence

On the basis of the occurrence, mushrooms may be classified into two groups:

1. **Epigenous** – Forming fruiting bodies entirely above the surface of substratum, *e.g.* cup fungi (*Peziza* species) and morels (*Morchella* species) (Figure 4).

2. **Hypogenous** - Such mushrooms usually grow underground and form fruiting bodies therein. For example: Some species of family Tuberaceae and these are also known as true truffles (Figure 5).

Amanita sp.

Agaricus sp.

Volvariella **sp.**

Marasmius sp.

Figure 3: The Presence or Absence of Annulus and Volva.

B) Natural Habitat

i) Humicolous – Humus inhabiting (Figure 6)
  a) Saprophytic – *e.g. Lepista nuda, Volvariella* species, *Marasmium* species, *Polyporus tuberaster*
  b) Symbiotic *e.g. Boletus* sp., *Lacterius* sp. and *Tricholoma* sp.

ii) Lignicolous- Wood inhabiting
  a) Saprophytic- *e.g. Agrocybe* sp., *Pleurotus* sp., *Auricularia* sp. and *Lentinus* sp.
  b) Parasitic *e.g. Armellariea mellea, Cyttaria* sp.

iii) Coprophilous – Dung inhabiting
  e.g. Agaricus species, *Coprinus* sp.

C) Colour of Spores

1. White spores
   Amanita, Armillaria, Cantharellus, Clitocybe, Lacterius, Lentinus, Lepiota, Pleurotus, Russula and *Tricholoma*

Cup fungi (*Peziza* sp.)

Morels

Figure 4: Epigenous Mushroom.

Truffles

Figure 5: Hypogenous Mushrooms.

Figure 6

I: Humicolous Mushroom

(a) Saprophytic

Lepista nuda   *Volvariella* **sp.**

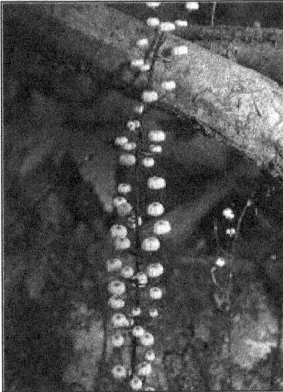

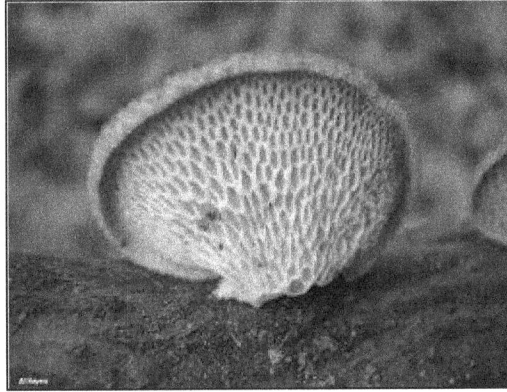

Marasmium **sp.**   *Polyporus tuberaster*

(b) Symbiotic

Boletus **sp.**   *Lactarius* **sp.**   *Tricholoma* **sp.**

Contd...

Figure 6–*Contd...*

II. Lignicolous Mushroom
Saprophytic

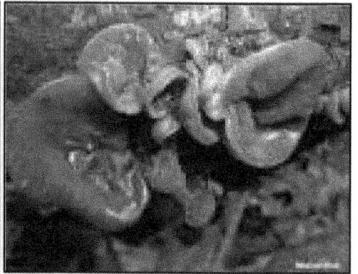

Agrocybe **sp.**   *Pleurotus* **sp.**   *Auricularia* **sp.**

Lentinus **sp.**

Parasitic

Armillaria   *Cyttaria* **sp.**

III. Coprophilous

Agaricus **sp.**   *Coprinus* **sp.**

2. Pink Spore
Flammulina, Inocybe and *Paxillus*

3. Black Spore
Entoloma and *Volvariella*

4. Purple Brown Spore
Agaricus, Hypholoma and *Stropharia*

5. Yellow Brown Spore
Coprinus, Panaeolus and *Psathyrella*

D) On the Basis of Morphology

I) Group – Basidiomycetous Mushrooms (Figure 7)

a) Fruiting Layers Exposed
This sub group has further been divided in to 5 categories.

1. **Gill Fungi:** Characterized by presence of gills, Gilled fungi have further been divided in to five categories according to the colour of the spores namely, (1) White coloured spores, (2) Pink coloured spores, (3) Brown coloured spores (4) Purple coloured spores and (5) Black coloured spores.
2. **Pore fungi** – Absence of gills and presence of tubers or pores.
3. **Teeth fungi** – Fungi without gills or pores but with prickles or teeth.
4. **Club or coral fungi** – Fungi without gills, pores or teeth, but are club or coral like in shape.
5. **Jelly fungi** – Fungi without gills, pores or teeth, forming soft, jelly like masses when moist.

b) Fruiting Layers not Exposed
1. Stem like receptacle arising from a definite egg stage, with volva at the base of the stipe. This category includes three important genera. These are *Dictyphora, Ithyphallus* and *Cynophallus*.
2. Stem like receptacle absent or rudimentary, volva absent. It includes five genera. These are *Geaster, Lycoperdon, Bovista, Calvatia* and *Cyathus*.

II. Group – Ascomycetous Mushrooms (Figure 8)
Morchella, Gyromitra, Helvella, Plicaria, Pustularia, Sarcoscypha and *Geopyxis*.

Shape of Fruit Bodies
i) **Gilled mushrooms** – (Family – Agaricaceae) *e.g. Agaricus* and *Amanita*.
ii) **Pore mushrooms** – (Family – Polyporaceae) *e.g. Fistulina, Trametes, Favolus candensis* and *Polyporus*.
iii) **Tooth fungal mushrooms** – (Family – Hydnaceae) *e.g. Hydnum corailloides* and *Hynum caputursi*.
iv) **Club shaped mushrooms** – (Family – Clavariaceae) *Clavaria*

I. Gill Fungi   **II. Pore Fungi**

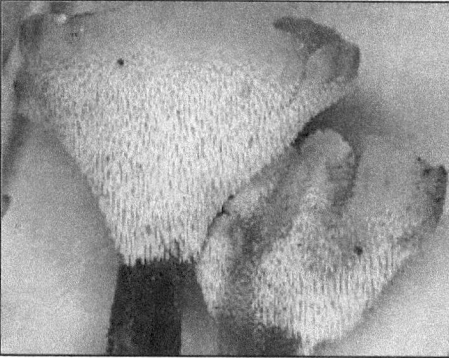 

III. Teeth Fungi   **IV. Club Fungi**

V. Jelly Fungi

Figure 7: Basidiomycetous Mushrooms.

Figure 8: Ascomycetous Mushrooms

I. Gilled Mushrooms

Agaricus **spp.**   *Amanita* **spp.**

II. Pore Mushrooms

a) *Fistulina*   **b)** *Trametes*

c) *Favolus canadensis*   **d)** *Polyporus*

Contd...

Figure 8–*Contd...*

III. Tooth Fungal Mushroom

a) *Hydnum coralloides*   b) *Hydnum caputursi*

V. Cup Shaped Mushroom

IV. Club Shaped Mushroom

Clavaria spp.   *Morchella* spp.

VI. Boletaceous Fungal Mushrooms

a) *Boletinus* spp.   b) *Boletus* spp.

Agaricus spp.

Coprinus spp.

Pleurotus spp.

Tricholoma spp.

Flammulina spp.

Calocybe spp.

Pholiota spp.

Kuehneromyces spp.

Hypholoma spp.

Stropharia spp.

**Figure 9: Mushroom of Subdivision: Basidiomycotina
Order: Agaricales.**

    v) Cup shaped mushrooms – *e.g. Morchella.*

    vi) Boletaceous fungal mushrooms – (Family – Boletaceae) *e.g. Boletinus* and *Boletus.*

Ainsworth *et al.* (1973) classified mushrooms into two subdivisions – Ascomycotina and Basidiomycotina. These two groups contain most of the edible fungi. Mushrooms belonging to these subdivisions may be described as under:

    i) Subdivision – Ascomycotina, class- Discomycetes, species *Morchella, Terfezia.*

    ii) Subdivisions – Basidiomycotina, class – Gasteromycetes.

There are six economically important orders of this class. These orders are Podaxals, Phallales, Sclerodermatales. Pyrenoganterales, Melanogastrales and Lycoperdales.

Ainsworth and Bisby's 'Dictionary of fungi' has been published during the year 1995 from International Mycolological Research Institute. According to this classification, the genera have been placed within families, all under 11 fungal phyla.

(I) Subdivision : Basidiomycotina

Class: Hymenomycetes

(a) Subclass : Holobasidiomycetidae
    Order : Agaricales (Figure 9)

    Genus :*Agaricus, Coprinus, Lentinus, Pleurotus, Tricholoma, Flammulina, Calocybe, Stropharia, Pholiota, Kuehneromycetes, Hypholoma.*

(b) Subclass: *Phragmobasidiomycidae*
    Order : *Auriculariales*, Tremellales (Figure 10)

    Genus : *Auricularia, Tremella*

Auricularia **spp.**   *Tremella* **spp.**

Figure 10: Mushroom of Order: Auriculariales and Tremella.

(II) Subdivision: Ascomycotina

Class : Discomycetes

Subclass : Hymenoascomycetidae

Order : Tuberales (Figure 11)

Genus : *Tuber*

Tuber spp.

Figure 11: Mushroom of Subdivision: Ascomycotina, Order: Tuberalles.

4

Edible Mushrooms

Edible mushrooms are the fleshy and edible fruit bodies of several species of macro fungi (fungi which bear fruiting structures that are large enough to be seen with the naked eye). They can appear either below ground (hypogeous) or above ground (epigeous) where they may be picked by hand. Edible mushrooms are consumed by humans as comestibles for their nutritional value and they are occasionally consumed for their supposed medicinal value. Mushrooms consumed by those practicing folk medicine are known as medicinal mushrooms. While hallucinogenic mushrooms (*e.g.* Psilocybin mushrooms) are occasionally consumed for recreational or religious purposes, they can produce severe nausea and disorientation, and are therefore not commonly considered edible mushrooms.

Edible mushrooms include many fungal species that are either harvested wild or cultivated. Easily cultivatable and common wild mushrooms are often available in markets, and those that are more difficult to obtain (such as the prized truffle and matsutake) may be collected on a smaller scale by private gatherers. Mushroom generally appears during rainy season from June to September. Some mushroom like Morchella spp. and Rhizopogon spp. appear during winter when the snow starts melting.

4.1. Common Edible Mushrooms: Commercially Exploited

1. *Armillaria mellea*

Cap 3-13 cm broad, convex at first, becoming nearly plane, sometimes broadly umbonate at maturity, margin finely striate; colour varying from yellowish-brown to

Figure 12: Common Edible Mushrooms (Commercially Exploited).

Armillaria mellea

Agaricus campestris

Auricularia auricula-judae

Boletus edulis

Cantharellus aurantiacus

Flammulina velutipes

Calocybe indica

Hericium coralloides

Hydnum repandum

Lactarius deceptivus

Lentinula edodes

Marasmius oreades

Contd...

Figure 12–*Contd...*

Morchella spp.

Pleurotus ostreatus

Pluteus cervinus

Tuber spp.

Tremella fuciformis

Volvariella volvacea

reddish-brown, the disc darker with fine hair or scales; surface viscid when moist; odour mild, taste acrid. Gills white to pallid, adnate to subdecurrent, sometimes notched, developing reddish-brown stains with age. Stipe 5-17 cm long, 0.5-3.0 cm thick, tapering towards the base when growing in clusters, enlarged to bulbous at the base when growing apart; tough, fibrous; partial veil cottony forming a superior ring; pallid above the ring, yellowish-brown to reddish-brown below. Spores 7-9.5 x 5-7 µm, smooth, elliptical, nonamyloid.

2. *Agaricus campestris*

Cap 3 to 10cm in diameter, the caps of field mushrooms is creamy white, sometimes developing small scales as they mature. Usually the margin remains down-turned or slightly in-rolled even when the cap has expanded fully. The thick flesh is white, sometimes turning slightly pink when cut but never staining yellow. Gills deep pink at first, the free crowded gills turn dark brown and eventually almost black as the fruitbody matures. Stem 3 to 10cm tall and 1 to 2cm in diameter, the white stem of Agaricus campestris is smooth above the single, delicate ring and somewhat scaly below. It is more or less parallel and does not turn yellow when cut. The ring itself is ephemeral, and by the time the fruitbody is fully developed there is rarely much evidence of a ring remaining. Spores ovoid, 7-8 x 4-5µm.

Other Species

A. bisporus, A.bitorquis.

3. *Auricularia auricula-judae*

The fruit body of *A. auricula-judae* is normally 3 to 8 centimeters across, but can be as much as 12 centimeters. It is distinctively shaped, typically being reminiscent of a floppy ear, though the fruit bodies can also be cup-shaped. It is normally attached to the substrate laterally and sometimes by a very short stalk. The species has a tough, gelatinous, elastic texture when fresh, but it dries hard and brittle. The outer surface is a bright reddish-tan-brown with a purplish hint, often covered in tiny, downy hairs of a grey colour. It can be smooth, as is typical of younger specimens, or undulating with folds and wrinkles. The colour becomes darker with age. The inner surface is a lighter grey-brown in colour and smooth. It is sometimes wrinkled, again with folds and wrinkles, and may have "veins", making it appear even more ear-like

Other Species

A. auricular, A. polytricha.

4. *Boletus edulis*

Cap with a slightly greasy penny-bun-like surface texture, the yellow-brown to reddish-brown caps of Boletus edulis range from 10 to 30 cm diameter at maturity. (An exceptionally large cap can weigh more than 1kg, with a stem of similar weight.) The margin is usually a lighter colour than the rest of the cap; and when cut, the cap flesh remains white, with no hint of bluing. Tubes and Pores: The tubes (seen when the cap is broken or sliced) are pale yellow or olive-brown and are easily removed from the cap; they end in very small white or yellowish pores. When cut or bruised, the pores and tubes of *Boletus edulis* do not change colour (as the pores of some otherwise quite similar species do). Stem a faint white net pattern (reticulum) is generally visible on the cream background of the stem, most noticeably near the apex. Clavate (club-shaped) or barrel-shaped, the stem of a Cap is 10 to 20cm tall and up to 10cm in diameter at its widest point. The stem flesh is white and solid. Spores subfusiform, smooth, 14-17 x 4.5-5.5µm.

Other Species

B.felleus, B.granulatus, B.luteus, B.retipes, B.sphaerosporus, B.subautreus.

5. *Cantharellus aurantiacus*

Cap 5-9 cm in diameter at first convex, later funnel shaped golden or orange yellow. Darker in the centre, margin curved, down or elevated. Flesh white, soft, gills 20-30 per cm, 3-5 mm wide, thicker at stem, forked, short decurrent, coloured like the cap. Stem 1.5-3 cm long, 1.5 cm thick at top, tapering upward, reddish brown.

Other Species

C. cibarius, C. infundibuliformis.

6. *Flammulina velutipes*

Cap 2 to 10 cm across and often distorted because of neighbouring caps in the cluster, the bright orange caps of Flammulina velutipes are generally somewhat darker towards the centre. Slimy in wet weather, the caps dry to a smooth sheen. Gills Adnate and broad, the gills of Velvet Shank fungi are white at first becoming pale yellow as the fruitbody matures. (The gills of cultivated forms of this mushroom usually remain white.) The stem is tough and covered in a fine velvety down. Usually pale near the cap, the stems often turn brown towards the base. Spores ellipsoidal, smooth, 6.5-10 x 3-4µm; inamyloid.

7. *Hericium coralloides*

Fruit body a mass of tufted stems, each with many pendant spines. Whole fungus may be 15-30 cm across and 20-40cm high; branches are white and stout and branch repeatedly from a central, basal point; tip of each branch has white spines 0.5-2cm long, in clusters like hands. Flesh firm; white. Odour pleasant. Taste pleasant. Spores ellipsoid, smooth or very finely roughened, 5-7 x 4, 5-6µ.

Other Species

H. erinaceous, H. lacinitum.

8. *Hydnum repandum*

Cap 2-12 cm broad, convex, becoming nearly plane, disc sometimes depressed; margin inrolled at first, lobed to undulate; surface dry, smooth to slightly scaly, cream to buff-orange, bruising to orange-brown; flesh thick, pale-buff, brittle, bruising buff-orange; odour and taste mild.Hymenophore teeth 0.4-0.6 cm long, brittle, cream-coloured, bruising orange-brown, sometimes decurrent. Stipe 2-7 cm tall, 1-2.5 cm thick, equal to enlarged at the base; attachment central to eccentric; surface dry, smooth, cream to buff-yellow, bruising orange-brown; veil absent. Spores 6.5-8.5 x 6-8 µm, nearly round, smooth. Spore print white.

9. *Lactarius deceptivus*

The cap is 7.5–25.5 cm in diameter, initially convex, but becomes funnel-shaped with age. The margin (cap edge) is rolled inwards and cottony when young, concealing the immature gills. The cap surface is dry, smooth and whitish when young, often with yellowish or brownish stains, but becomes coarsely scaly and darkens to dull brownish-ochre with age. The gills have an adnate to decurrent attachment to the stem, close to subdistant, white at first then cream to pale ochre. They are 5–65 mm long and 1–7 mm deep. The stem is 4–10 cm long, up to 3 cm thick, nearly equal in width throughout or tapered downward. It is dry, scurfy to nearly smooth and white, staining brown with age. It is initially stuffed (as if filled with cotton), but later becomes hollow. The flesh is thick and white, and between 3 and 15 mm thick. The latex produced by the mushroom is white, and does not change colour upon exposure to air, although it stains the mushroom flesh a yellowish-brown colour.

Other Species

L. deliociosa, L. indigo, L. piperatus, L. subdulcis, L. torminosus, L. trivialis, L. vellerius, L. volemus.

10. *Lentinula edodes*

Cap 5-25 cm broad, black when young, dark brown to light brown with age and hemispheric, expanding to convex and planar at maturity. Spores 5-6.5 x 3-3.5µm, ovoid to oblong ellipsoid. Brown Gills white and even at first, serrated with age. Stipe fibrous and tough. Veil absent.

11. *Marasmius oreades*

Cap 1.5-4.0 cm broad, rounded, becoming campanulate (bell- shaped), finally convex to nearly plane, typically with a broad umbo; margin striate when moist, sometimes upturned with age; surface dry, smooth, light cinnamon-brown to buff-brown when young, fading with age to pale-buff or cream; flesh thick, reviving after being dried; odour and taste mild. Gills adnexed to nearly free, subdistant, interspersed with shorter lamellae and intervenose; cream to buff-coloured. Stipe 2-6 cm tall, 2-5 mm thick, equal, tough, cream to buff above, the base usually somewhat darker and pruinose; veil absent. Spore 7-8.5 x 4-5.5 µm, elliptical, inequalateral, apiculate (pointed at the end), smooth, nonamyloid; spore print white to pale-cream.

Other Species

M. prasiosmus, M. erythropus, M. scorodonius.

12. *Morchella* spp.

Cap more or less conical, irregular, cylindrical or narrow conical, spherical or almost spherical, ridges and pits often extending longitudinally or ridges irregularly branched. Pits large and hallow or rounded or irregular. Stem 4-10 cm long, 2-3 cm thick, white or pale tan, hollow, brittle.

Other Species

M. conica, M. crassipes, M. esculenta.

13. *Pleurotus ostreatus*

The mushroom has a broad, fan or oyster-shaped cap spanning 5–25 cm; natural specimens range from white to gray or tan to dark-brown; the margin is inrolled when young, and is smooth and often somewhat lobed or wavy. The flesh is white, firm, and varies in thickness due to stipe arrangement. The gills of the mushroom are white to cream, and descend on the stalk if present. If so, the stipe is off-center with a lateral attachment to wood. The spore print of the mushroom is white to lilac-gray, and best viewed on dark background. The mushroom's stipe is often absent. When present, it is short and thick.

Other Species

P. sapidus, P. sajor-caju, P. florida, P. eryngii.

14. *Pluteus cervinus*

The cap can be up to 15 cm in diameter, but is often much smaller. Initially it is bell-shaped, and often wrinkled when young. Later it expands to a convex shape. The cap can be deer-brown, but vary from light ochre-brown to dark brown, with a variable admixture of grey or black. The centre of the cap may be darker. The cap surface is smooth and matte to silky-reflective. The cap skin shows dark radial fibres when seen through a lens, indicating that the microscopic cuticle structure is filamentous. The gills are initially white, but soon show a distinctive pinkish sheen, caused by the ripening spores. The stipe is 5-12 cm long and 0.7-2.0 cm in diameter, usually thicker at the base. It is covered with brown vertical fibrils on a white ground. The flesh is soft and white. The mushroom has a mild to earthy radish smell and a mild taste at first, which may become slightly bitter. The spore size is approximately 8×5μ, and the individual spores are elliptical and smooth. The spore print is salmon-pink to reddish-brown.

15. *Tuber* spp.

The fruit body is irregularly globose, fleshy or leathery, with a smooth or warted peridium. The interior has two kinds of veins, which frequently gives it a marbled appearance. The asci are globose, ovoid and contain usually one to four coloured, reticulate or spiny, ellipsoid or spherical, large spores.

Other Species

T.magnatum, T.rufum.

16. *Tremella fuciformis*

Fruit bodies are gelatinous, watery white, up to 7.5 cm across (larger in cultivated specimens), and composed of thin but erect, seaweed-like, branching fronds, often crisped at the edges. Microscopically, the hyphae are clamped and occur in a dense gelatinous matrix. Haustorial cells arise on the hyphae, producing filaments that attach to and penetrate the hyphae of the host. The basidia are tremelloid (ellipsoid, with oblique to vertical septa), 10–13 x 6.5–10 μm, sometimes stalked. The basidiospores are ellipsoid, smooth, 5–8 x 4–6 μm, and germinate by hyphal tube or by yeast cells.

Other Species

T. aurantia.

17. *Volvariella volvacea*

Cap 5-16 cm; egg-shaped when young, expanding to convex or broadly conic, becoming broadly convex or nearly flat; dry; radially streaked with hairs; gray to brownish gray or grayish brown or nearly black when young, with a paler marginal area; soft; the margin not lined, but often splitting with age. Gills free from the stem; white becoming pink; close or nearly crowded. Stem 4-14 cm long; up to 2 cm thick; tapering gradually to apex, with a swollen base; dry; whitish or brownish; silky; the base encased in a thick, sack-like volva that is brownish gray to nearly black above and whitish below.

18. *Calocybe indica*

Calocybe indica have a milky white colour with a cap that can grow as large as 15 cm in diameter and a stem which may exceed 30 cm in length. One mushroom can weigh as much as 2kg – under optimal environmental conditions and using suitable substrate. Cap may be Campanulate -*i.e.* bell shaped or convex in the initial stages of development and flattens as it grows further and becomes ready to be harvested. Gills are sinuate or smoothly attached and run down the stem. Stalk long, thick fibrous. Spore print is white. The size and shape of the cap during cultivation, depends mainly on the light and the oxygen availability as well as the amount of substrate available.

4.2. Non Exploited Edible Mushrooms (Figure 13)

1. *Amanita rubescens*

Cap 4-15 cm; convex, becoming broadly convex or flat; dry or slightly sticky; adorned with numerous felty warts that are bright yellow and densely packed at first, but soon spread and fade, becoming pinkish, grayish or dull tan; surface dull brassy yellow to dull brown when young, becoming flushed with red shades, and eventually turning reddish brown to tan or brown; the margin typically not lined. Gills are free from the stem or narrowly attached to it; white, sometimes discolouring reddish; close or crowded; short-gills numerous. Stem are 5-18 cm long; 1-3 cm thick; more or less equal, or sometimes slightly enlarged toward base; the base indistinct to bulbous; generally without universal veil remnants; without a rim; white at first, becoming stained pinkish to dirty red; bald or finely hairy; with a fragile, persistent ring. Flesh is white throughout, discolouring slowly pale pinkish red, especially around worm holes.

Other Species

A. solitaria, A. spissa.

2. *Bovista plumbea*

Fruiting body 1.5-3.5 cm broad, globose to slightly compressed, attached to the substrate by a tuft of mycelium; exoperidium white, glabrous, becoming buff to pale-tan and minutely tomentose, sometimes areolate; exoperidium flaking away, or peeling off in sheets, the latter typical of maturation in hot, dry conditions; endoperidium membranous, lead-grey, with or without adhering fragments of exoperidium; spores released via a small apical pore; gleba white, turning dingy yellowish, olive-brown, finally dark-brown, firm-textured; subgleba and sterile base absent. Spores 5.0-6.5 x 4.0-5.5 µm, ovoid, thick-walled, nearly smooth, with a central oil droplet, and a 7.5-11.5 µm pedicel; capillitium of individual elements, not interwoven, main branches thick-walled, flexuous, rapidly tapering, forking more or less dichotomously.

3. *Clavaria pistillaris*

Occasionally somewhat laterally flattened and longitudinally wrinkled or grooved, these large simple (not forking) truncheon-shaped clubs (tapering slightly towards the base) have rounded tips and are at first yellow, turning various shades of pink, mauve, violet and brown with age or when bruised. The individual clubs are typically 8 to 30 cm tall and 5 to 8 cm across at their widest point when fully developed. Firm when young, the white flesh of *Clavariadelphus pistillaris* turns violet-brown when cut; it becomes soft and spongy when fruitbodies reach full maturity. Occasionally the hollow upper region of an old fruitbody splits, and insects are then able to access the interior via holes in the top of the club. This should sound a warning note to anyone intent on gathering for the pot these reportedly edible but far from delectable fungi. Spores ellipsoidal, smooth, 11-16 x 6-10µm, non amyloid with an eccentric germ pore.

Other Species

C. amethystine, C. flava, C. stricta.

Figure 13: Non Exploited Edible Mushrooms.

Amanita rubescens

Bovista plumbea

Clavaria pistillaris

Calvatia utriformis

Clitocybe candida

Coprinus comatus

Clitopilus prunulus

Dictyophora duplicata

Lepiota cristata

Lycoperdon perlatum

Melanogaster variegatus

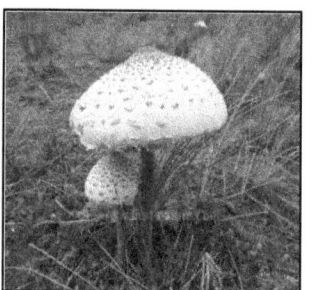

Macrolepiota procera

Contd...

Figure 13–*Contd...*

Podaxis pistillaris

Pisolithus arrhizus

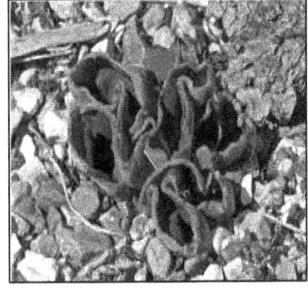

Peziza cochleata

Pholiota squarrosa

Psilocybe aeruginosa

Rhizopogon luteolus

Russula aurata

Stropharia aeruginosa

Tricholoma melaleucum

Terfezia **spp.**

Verpa bohemica

4. *Calvatia utriformis*

Fruit body 5-12 cm wide, 6-14 cm tall, spherical top with a cylindrical or tapering stalk. Depressed above with a stout thick base and cord like root. Whitish at first, yellow or brown when old, surface covered with warty patches, spines or cracks.

Other Species

C. craniformis, C. cyathiformis, C. maxima.

5. *Clitocybe candida*

The cap is usually fleshy, plain depressed or infundibuliform with the margin involute, 10-30 cm wide, white to very pale tan, smooth. Flesh 1-2.5cm thick at the stem, white, firm. The gills 4-7 mm wide, typically decurrent sometimes acutely adnate, but never sinuate. The stem is 1.5-4 cm thick, 6-10 cm long, fibrous externally elastic, spongy, stuffed.

Other Species

C. infundibuliformis, C. laccata, C.maxima, C. multiceps.

6. *Coprinus comatus*

The shaggy ink cap is easily recognizable from its almost cylindical cap which initially covers most of its stem. The cap is mostly white with shaggy scales, which are more pale brown at the apex. The free gills change rapidly from white to pink, then to black. It is deliquescent. The stipe has a loose ring and measures 10–37 cm high by 1–2.5 cm diameter. Microscopically it lacks pleurocystidia. The spore print is black-brown and the spores measure 10–13 × 6.5–8 μm. The flesh is white and the taste mild.

Other Species

C. micaceus, C.atramentarius, C. sterquilinus.

7. *Clitopilus prunulus*

The cap is initially convex when young, but in maturity flattens out, usually with a shallow central depression. It is white or light grey or yellow, sticky when moist, and 3 to 10 cm in diameter with a characteristic feel to the touch of chamois skin. The gills are decurrent in attachment to the stipe, spaced together rather closely, and whitish, although they often develop a pinkish hue with age. The stipe is 3 to 8 cm long x 4–15 mm thick, and white. This mushroom has a mealy odour, somewhat like cucumber. The spore print is pink. Spores are 9–12 x 5–6.5 μm. C. prunulus may be found growing on the ground in hardwood and coniferous woods in the summer and fall. The variant C. prunulus var. orcellus has a slimy cap and white colours.

Other Species

C. abortivus, C. orcella.

8. *Dictyophora duplicata*

Stalk 10-15 cm long, 3-4 cm thick cylindrical, tapering at the tip and the base white, hollow with an opening at the tip into the hollow, cap conical to almost globose, 3-5 cm high, surface of the cap covered with angular, shallow, thin walled chambers each 4 mm in diameter, covered with olive brown evil smelling slime, joined to the top of the stem by a rounded collar. A net like veil is attached to the tip of the stalk beneath the cap and flares out below the lower edge of the cap for 3-4 cm.

9. *Lepiota cristata*

Cap 1.5-4.5 cm broad, convex, nearly plane with age, often umbonate; surface at first smooth, reddish-brown at the disc, shading to a lighter margin, the cuticle, except at the disc, soon breaking up into concentrically arranged scales revealing underlying white tissue; flesh thin, white; odour pungent. Gills free, close, white to cream. Stipe 2.5-6 cm tall, 0.2-0.4 cm thick, hollow, slender, equal to slightly enlarged at the base; surface white, smooth or with a scattering of fibrils; veil membranous forming a fragile, superior ring. Spores 6-7.5 x 3-4.5 µm, wedge-shaped, spurred, smooth and dextrinoid; spore print white.

Other Species

L. rachodes, L. naucina.

10. *Lycoperdon perlatum*

Fruit body 2-9 cm tall, 2-4 cm broad, subglobose to pyriforme to elongated pyriforme, at maturity with an apical pore for spore dispersal; ground colour white to pallid to pale brown; surface composed of conical spines, 1-2 mm tall, surrounded by a persistent circular row of warts; spines white to pallid to pale brown, leaving scars as they fall off. Base well developed, forming a pseudostipe; spines and warts absent or much less prominent. Gleba white, firm; becoming soft and yellow, then brown to dark brown and powdery as spores mature. Spores 3-4.5 µm, globose, verrucose. Spores in deposit yellow-brown to olive-brown to dark brown.

Other Species

L. pyriformin, L. peckii, L. pulcherrimum.

11. *Melanogaster variegatus*

Fruit body ochraceous or clear yellow then reddish, irregularly globose or nodular tomentose and ornamented with brown fibrous cord like anastomosing mycelium. Gleba fuliginous, then black, tramal plates whitish, then bright orange.

Other Species

M. ambigus, M.inquinans.

12. *Macrolepiota procera*

Cap initially spherical and pale brown with a darker brown area near the crown that breaks into scales, the cap of *Macrolepiota procera* expands until it is flat with a small central bump, known as an umbo. The cap flesh is white and does not change significantly when cut. The cap diameter at maturity ranges between 10 and 25 cm. Gills broad, crowded gills of the Parasol Mushroom are white or pale cream and free, terminating some distance from the stipe. Stem large, double-edged ring persists around the stem of *Macrolepiota procera* but often becomes movable and falls to the base. The stem is smooth and white or cream but decorated with small brown scales that often give it a banded, snake skin appearance. Inside the stem the tough white fibrous flesh is loosely packed, and sometimes the stem is hollow. Bulbous at the base, the stems of *Macrolepiota procera* tapers inwards slightly towards the apex; the diameter ranges from 1 to 1.5 cm (to 2.5 cm across the bulbous base), and the stem height can be up to 30 cm. Spores ellipsoidal, smooth, thick-walled; 12-18 x 8-12μm; with a small germ pore.

13. *Podaxis pistillaris*

Cap narrowly ellipsoid to obovate, 4.0-11.0 cm tall, 2.0-4.0 cm broad, not expanding, the margin fused to stipe except occasionally breaking free in senescence; surface two-layered, the outer layer cream-white, sparsely covered with flattened tan-coloured scales, partially sloughing off with age, revealing a thin, brittle, glabrous cream to buff-coloured inner layer, the latter eventually fissured or cracked, releasing spores; odour and taste untried. Gills replaced by a puffball-like gleba, at first pallid to yellowish, sometimes with vinaceous tints, at maturity dark-brown, dark reddish-brown to blackish, the spore mass embedded in a capillitium. Stipe 4.0-9.0 cm long, 0.5-1.0 thick, straight, more or less equal except for a soil-encrusting, abruptly bulbous base; surface covered by a whitish to cream-coloured membrane, fragmenting with age, exposing a pallid, longitudinally striate under-surface; context consisting of a woody cortex and cottony-fibrous pith, lacking the "cottony cord" seen in *Coprinus comatus*; partial veil absent. Spores 10-15.0 x 9.5-13.5 μm, globose to ovoid, smooth, with a thick double-wall, prominent germ pore, inconspicuous hilar appendage, and lacking a pedicel.

14. *Pisolithus arrhizus*

The fruit body is subglobose, brown, sometimes with a well defined stem like base, sunk into the ground and attached to thick yellow mycelial cords, smooth or subtuberculose, peridiola sulphur-yellow then brown numerous, oval or irregularly angular by pressure, slightly hairy.

15. *Peziza cochleata*

Disc fulvous to deep brown, externally dingy ochraceous to white scurfy, globose, then expanded becoming bent with involute margin, sessile, usually split, laterally to the tomentose base, variously contorted and plicate.

16. *Pholiota squarrosa*

The cap ranges from 3 to 12 cm in diameter, and depending on its age, can range in shape from bell-shaped to rounded to somewhat flattened. The cap colour is yellowish-brown to tawny in older specimens. The scales on the cap are yellowish to tawny, and recurved. The stem is 4 to 12 cm long by 0.5 to 1.5 cm thick, and roughly equal in width throughout. The partial veil that covers the young gills forms a thick, woolly ring on the upper part of the stem. Above the level of the ring, the stem is bare, while below it is scaly like the cap. The gills are covered by a partial veil when young and have a greenish-brown colour; mature gills are rusty brown. They are crowded closely together, attached to the stem (adnate), and usually notched (sinuate). The spore print is cinnamon or rusty brown. The spores are elliptic, smooth-walled, nonamyloid (not absorbing iodine when stained with Melzer's reagent), and measure 6.6–8 by 3.7–4.4 µm. The basidia (spore-bearing cells) are club-shaped, and four-spored, with dimensions of 16–25 by 5–7 µm.

Other Species
P. adiposa, P. mutbilis, P, praecox.

17. *Psilocybe aeruginosa*

The cap is 1–6 cm in diameter, conical to convex, tan brown, hygrophanous, margin striate when moist, uplifting with age, and often with a slight umbo. It bruises bluish where damaged. The gills are crowded, cream colour when young, violet brown with age, with adnate to adnexed attachment. The spore print is dark violet brown. The stipe is 4.5 to 22 cm long, 0.2 to 0.5 cm thick, white, bruising blue where damaged, finely striate, and equal to slightly enlarged near the base. A white cortinate partial veil soon disappears and often leaves traces in the upper stem. The taste and odour are farinaceous. The cheilocystidia are 17 - 29 x 5.5 - 11, hyaline, fusoid-ventricose, subpyriform or mucronate, often with an elongated neck at the apex which is 2 - 4.5 µm. The pleurocystidia measure 22 - 47 x 6 - 16 µm and is shaped like the cheilocystidia and also hyaline. The spores are smooth, subellipsoid, with an apical germ pore, measuring 13.2 - 14.3 x 6.6 - 7.7 x 6 - 7.5 µm.

Other Species
P. foenisecii, P. spadicea.

18. *Rhizopogon luteolus*

Looking very much like a potato and just as variable in size and shape, the Yellow False Truffle is typically 1.5 to 4.5 cm across its largest dimension and can be ovoid, ellipsoidal, an oblate spheroid. There is no stem, but cord-like mycelial threads spread into the soil and to tree rootlets from a central point under the fruitbody. Its outer skin is tougher than the interior tissue, and it is off-white initially but soon turns ochre and eventually olive brown. The outer surface, which usually cracks irregularly as the fruitbody expands, is often randomly decorated with tawny mycelial strands that give it a slightly woolly appearance. Internally the spore-bearing gleba of *Rhizopogon luteolus* is almost white at first turning ochre and eventually olive brown as the spores approach maturity. The interior of the fruitbody comprises many tiny chambers lined with basidia, upon which the spores develop; initially the interior is soft and spongy, becoming dry and powdery when old. Spores oblong-ellipsoidal, 7-10 x 2.5-3.5µm; covered with an irregular coarse reticulum. Spore mass creamy white or yellowish.

Other Species
R. rubescens.

19. *Russula aurata*

The cap is 4–9 cm (1.6–3.6 in) wide and a blood- or orange-red in colour with ridged margins. Sticky when wet, it is initially convex, then later flat, or depressed; it is able to be peeled half-way. The widely spaced gills are ochre with yellow edges, and adnexed or free. The stipe is up to 3–8 cm long and 1–2.5 cm wide, cylindrical and white or cream to golden-yellow. The brittle flesh is yellow and the taste mild. The spore print is ochre, the warty spores are oval or round and measure 7.5–9 × 6–8 µm.

20. *Stropharia aeruginosa*

The cap is convex, broadening, and becoming umbonate with age. It is from 2–8 cm in diameter. At first it is a vivid blue/green, and very glutinous (slimy), with a sprinkling of white veil remnants around the edge. The colour in the gluten fades, or is washed off as it matures, and it becomes yellow ocher, sometimes in patches, but mostly at the centre. Finally, it will lose the blue-green colouration completely. The stem is quite long and of uniform thickness. It has a fragile brown/black ring, and below this the stem is covered in fine white scales, or flakes. The gills are initially white, then clay-brown, and sometimes have a white edge. The spore print is brownish-purple, and the oval spores 7–10 x 5 µm.

Other Species

S. semiglobata.

21. *Tricholoma melaleucum*

Smoky brown when moist, grayish brown when dry, cap 2-7cm wide, slightly convex with umbo, flesh 2-7 mm thick near the stem disappearing towards margin. Gills 20-25 per cm sinuate or almost free white at first, later pale brown. Stipe streaked with smoky fibrils, base fuscous.

Other Species

T. magnivillarae, T. personatium, T. terreum, T.equestre.

22. *Terfezia* spp.

Fruit body pear shaped. Asci big and ascospore globose.

23. *Verpa bohemica*

The cap of this fungus (known technically as an apothecium) is 2 to 4 cm in diameter by 2 to 5 cm long, with a conical or bell shape. It is folded into longitudinal ridges that often fuse together (anastomose) in a vein-like network. The cap is attached to the stem at the top only—hanging from the top of the stipe, with the lobed edge free from the stem—and varies in colour from yellowish-brown to reddish-brown; the underside of the cap is pale. The stem is 6 to 12 cm long and 1 to 2.5 cm thick, cream-white in colour, and tapers upward so that the stem is thicker at the base than at the top. Although the stem is initially loosely stuffed with cottony hyphae, it eventually becomes hollow in maturity; overall, the mushroom is rather fragile. The spore deposit is yellow, and the flesh is white.

5

Poisonous Mushrooms

The mushrooms growing in nature are known as wild mushrooms, and many of them are poisonous which includes probably more than 200 species. Many mushrooms formerly considered doubtful or poisonous have been found to be edible. The original misconception in these cases probably resulted from observation of sickness following the consumption of mushrooms that were no longer fresh and putrefied due to improper storage. The tragic consequences which may sometimes arise from the accidental partaking of poisonous mushrooms were known to the ancients of Occident as well as Orient. Number of cases of mushroom poisoning has been referred to in classical Greek and Roman literature. Mushroom lack chlorophyll; grow saprophytically on various substrates or parasitically and symbiotically on or with other organisms. They have been attracting attention of mankind since ancient times. The use of mushroom as food is as old as human civilization. However, all mushrooms are not edible. Some are mild to deadly poisonous. But there is no visible sign on any mushroom nor there is any thumb rule which indicates whether it is poisonous or not. In older days, before people could differentiate between edible and poisonous mushrooms, many lives were perhaps lost by eating poisonous mushrooms. There is no general rule for identification of poisonous mushrooms. Mushroom poisoning mostly occurs in rainy month, when people collect wild mushrooms. Identifications and isolation of toxins which are the active principles of mushroom poisoning is very difficult and time consuming.

Some of the traditional methods for identification of edible and non edible/poisonous species are unreliable. It was believed that edible fungus peeled off easily and did not change a silver spoon black while cooking. This belief is incorrect as *Amanita phalloides* (death cap), which is poisonous, peeled off easily and the silver spoon remains unaffected. The belief was prevalent that brightly coloured mushrooms were poisonous, where as white or creamy ones were edible. But this is also incorrect as Chanterelle and Woodblewits are bright coloured, but are quite safe to eat; whereas, "Death cap". "Fools mushroom" and Destroying Angel are white coloured and poisonous. Not only the Greek and Roman writers put forth different views about fungi; many great mycologists since the dawn of mycology have given classification and taxonomical characters of fungi in their treatise. It is not safe to believe that mushroom from one genus are mostly edible as in case of *Agaricus. A. bisporus* is edible while *A. xanthodermus* is poisonous. Mushroom which are morphologically similar or have close resemblance would not include all edible or non edible type. *Lepiota margani* and *L. rachodes* look alike when half grown however, *L.margani* is poisonous and *L. rachodes* is edible and delicious. Some mushroom are poisonous when consumed in large quantity *e.g. Verpabohemica*, while some are edible while fresh, but old decayed are poisonous *e.g. Armillariella mella*.

5.1. What is Mushroom Poisoning

Poisoning by consumption of wild or unknown mushrooms is one of the most common types of food poisoning due to mushroom.

It occurs mainly in people moving into the woods during the weekend to pick mushrooms and eating their recipe. It is a kind of intoxication caused by unknown mushrooms and their toxins that affects both men and women.

5.1.1. The General Symptoms

The first symptoms of mushroom poisioning develops after 8-12 hours of use and are practically independent of the type of mushroom often with a sign of acute gastroenteritis. Initially, there is spastic pain throughout the abdomen, nausea, repeated vomiting, then loose stools, weakness, pale skin and cool extremities.

The most serious is the poisoning death cup: eating of ¼ of the mushroom cap can cause the victim to his death. In cases of poisoning by this fungus the vomiting resemble coffee grounds, gastric bleeding in the feces, with admixture of blood and stool frequently in about 20-25 times a day.

The reaction of the nervous system to different types of toxic substances in the mushrooms is anxiety, restlessness, hallucinations and motor stimulation.

With the progression of mushrooms poisoning, there are signs of oppression heart activity, decreasing heart rate and blood pressure, dizziness and headache, skin and mucous membranes become blue, and extremities become cold and wet.

Severe consequences of mushroom poisoning, especially death cup are associated with toxic effects on the liver and kidneys. After 8–12 hours of poisoning cells begin to break down, after 2-3 days, these changes become irreversible, developed into acute hepatic and renal failure. Fungal toxins affect all organs and systems, and if, during the first 2-3 days without urgent medication the person is bound to die. It is difficult to cope with the effects of mushroom poisoning in children and the elderly.

5.1.2. What to do: The First Aid

As soon as the sign of mushrooms poisoning appears one should immediately call "Ambulance" or doctor.

Before the arrival of the doctor one should take the following measures:

☆ Stomach wash: Drink 5-6 glasses of water, and then by pressing a finger or spoon on the tongue, induce vomiting. The procedure is repeated 3-5 times.

☆ In the absence of diarrhea one should take a laxative, for example, 1 tbsp of castor oil or petroleum jelly, or 30 ml of a 33 per cent solution of magnesium sulfate.

What Not to Do

When there are signs of mushroom poisoning taking of food and drinking alcohols is strictly forbidden, one should not take medicines for pain, fever, diarrhea and vomiting.

Remember that only the timely medical assistance will help to cope up with the effects of mushrooms poisoning. The human body is not able to neutralize the mushroom poisons and lost time can be fatal.

http://prohealthnews.Net/health/mushroom.poisoning-first signs and first aid.html.

5.2. Clinical Symptomatology and Management of Mushroom Poisioning

Koppel (1993) described the clinical symptomology and management of mushroom poisoning. Among poisonous mushrooms, a small number may cause serious intoxication and even fatalities in human being. Humans may become symptomatic after a mushroom meal for rather different reason:

1. Ingestion of mushrooms containing toxins,
2. Large amounts of mushrooms may be hard to digest,
3. Immunological reactions to mushroom-derived antigens,
4. Ingestion of mushrooms causing ethanol intolerance, and
5. Vegetative symptoms may occur whenever a patient realizes that there might be a possibility of ingestion of a toxic mushroom after a mushroom meal.

Based on the classes of toxins and their clinical symptoms, seven different types of mushroom poisoning can be distinguished:

1. Phalloides
2. Orellanus
3. Gyromitra
4. Muscarine
5. Pantherina

6. Psilocybin
7. Gastrointestinal mushroom syndrome.

Two other entitles of adverse reactions to mushrooms are

8. Coprinus
9. Paxillus syndrome.

Phalloides, orellanus, gyromitra and paxillus syndrome may lead to serious poisoning, which generally requires treatment of the patient in an intensive care unit. Diagnosis of mushroom poisoing is primarily based on anamnestic data, identification of mushrooms from leftovers of the mushroom meal, spore analysis, and/or chemical analysis.

There are two types of intoxication

1. Clinically Recoverable (Mild) Mushroom Poisoning

It usually occurs within half an hour to four hours after consuming the poisonous/ toxic mushrooms. Toxicity in this case is mild and the mushrooms that produce these poisoning are :

a. Mushrooms of the Genera *Clitocybe, Russula, Boletus, Agaricus* and *Lactarius*

It produces gastrointestinal poisoning. It is due to the the action of certain emetic fatty acids, such as agaric, lauric.

The symptoms appear from a quarter of an hour to four hours after eating such mushroom. The main symptoms are vomiting, diarrhea, dizziness, or headache. The treatment involves the emptying of the stomach and the application of body fluids to avoid dehydration.

b. Mushrooms of the Genera *Conocybe, Psilocybe, Paneolus,* Pluteus, Copelandiam, Stropharia or Inocybe

They produce hallucinations due to their contain of norbaeocystin, baeocystin, psilocin and pasilocybin.

The symptoms appear 15 minutes after eating mushroom and before two hours. The main symptoms are vomiting, nausea, hypotension, headache, amnesia and especially hallucinations, usually resolving spontaneously after 4 or 10 hours.

c. *Amanita muscaria* and *Amanita pantherina*

They produce micoatropinic or panterianic nervous disorders, because they mainly contain muscaridin and other principles such as neurin, choline, isomuscarine, ibotenic acid, pyro-ibotenic acid, muscimol or muscazone

The main symptoms take place within half an hour to three hours after eating the mushrooms. The main symptoms include diarrhea, vomiting and dry mouth.

After the first described symptoms, hallucinations can take place, especially of a visual type. Mental confusion, convulsions, dilated pupils and coma can also appear. The treatment involves intestinal emptying, the use of sedatives and hydration of the body.

d. Mushroom of the Genus *Clitocybe* and *Inocybe*

They produce sweating disturbances that start between 15 minutes and four hours after eating the mushroom. It is produced by the action of muscarine, betanin, neurin, isomuscarina and choline. It is mainly characterized by the onset of vomiting, diarrhea, weakness, lack of urination, profuse sweating, decreased heart rate, constriction of pupils, anxiety, etc. Treatment involves gastric lavage and hydration.

e. Mushrooms of the genus *Coprinus* (*C. atramentarius, C. africanus, C. insignis, C. erethisteus, C. micaceus*) and *Clitocive* (*C. clavipides*)

They produce coprinic disorders, because they contain a principle similar to disulfiram that prevents the metabolism of acetaldehyde. It is toxic only when combined with alcohol (if you drink alcohol 3 or 4 hours before eating the mushroom or two or three days after eating them)

Symptoms occur after a quarter of an hour or half an hour after drinking alcohol. The main symptoms are nausea, vomiting, very strong hot feeling in the head and neck, arrhythmias, hypotension.

The problem usually resolves spontaneously within three to six hours. Treatment may involve the use of drugs to treat hypotension and arrhythmia.

2. Fatal (Major) Mushroom Poisoning

a. Mushrooms of the Genus *Amanita* (*A. phalloides, A. verna, A. porrinensis* and some Species of *Lepiota* and *Galerina*)

They produce phalloid disorders. The main components of these mushrooms are amaninamide, amanine, alpha, beta and gamma amanitin, aloidine, phalloidin, phalloin, phallisacin, phallacidin, phallacin, viroidine, etc.

Symptoms of poisoning occur after about 8 or 10 hours after eating of mushroom or within 24 hours. The most characteristic symptoms are severe diarrhea, vomiting, excessive thirst, sweating, cramps, stomach pain, headache, dehydration, etc.

Later, after an apparent recovery, a few days after having ingested them, a number of symptoms that indicate damage to the liver and kidney occur : lack of urination, jaundice, bleeding, pulmonary edema, hypoglycaemia, etc. It can cause problems of lack of awareness and behavioural disorders.

The final episode involves limb paralysis, convulsions, coma and death.

Only proper medical treatment in the first hours after ingestion may save the person who has eaten these mushrooms. The treatment is very complex and may even need a liver transplant.

b. Mushrooms of Genus *Cortinarius* (mainly *C. orellanus*)

They produce cortinaric disturbances. The first symptoms occur after 3 or 15 days after eating mushroom.

The most characteristic symptoms are mild belly pain, intense thirst, headache, muscle pain, cramps, nephritis, kidney failure, liver failure, hyperglycemia, uremia, pulmonary edema, and so on.

The treatment involves the use of appropriate medications to protect the liver and kidneys. The use of hemodialysis has led to the recovery of most affected patients.

c. Mushrooms of the Genus *Gyromitra* (*G. esculenta, G. gigas, G. ingula*) and *Morchella* (*M. esculenta*)

They produce gyromitric poisoning. They contain a chemical compound called gyromitrins, which are very harmful to the liver and kidneys.

Symptoms usually appear after about 6 or 9 hours after eating mushroom but can occur after almost a whole day. The main symptoms are vomiting, diarrhea, arrhythmias and hypotension. After one or two days of poisoning, symptoms may become more severe with enlarged liver, elevated transaminases, destruction of red blood cells, black urine etc.

The treatment of this poisoning is gastric lavage and intestinal protection of the liver and affected kidneys.

There is no home remedy to nullify or cure the toxicity of mushrooms. Therefore, the only option is to abstain from eating wild/unknown mushrooms. The best way to avoid mushroom poisoning is awareness and prevention. One should only pick those mushrooms one can identify very well. In case of doubt, it would be better not touch an unknown mushroom.

In case of suspicion of having eaten a toxic mushroom, one should seek prompt medical attention. The more rapid the treatment for mushrooms poisoning, the less effect of this poisoning will have on body organs.

http://www.botanical-online.com/english/mushroom poisoning.html.

Horowitz and Hendrickson (2013) proposed approach consideration and supportive measures to treat mushroom toxicity.

Approach Consideration

In the absence of a definitive identification of the mushroom, all ingestions should be considered serious and possibly lethal. Once mushroom toxicity is diagnosed, treatment is largely supportive. Early volume resuscitation is important for liver and renal toxic syndromes.

Gut decontamination including whole-bowel irrigation may be necessary for amatoxins. Beyond the first postprandial hour, orogastric lavage is not recommended, because of its questionable efficacy. Activated charcoal plays a much more important role in limiting absorption of most toxins and is indicated for all patients with amatoxin mushroom poisoning, regardless of the timing of presentation. When amatoxins are suspected, multiple doses of activated charcoal should be administered repeatedly to interrupt enterohepatic circulation of these toxins.

In general, children are more susceptible to volume depletion and mushroom toxicity (mushroom poisoning) than are healthy adults. Elderly patients are also more susceptible to volume depletion than are healthy adults.

Once a toxin is absorbed, it may potentially be neutralized by inhibition of tissue uptake of the toxin, inhibition of the metabolic pathways involved in the development of toxicity and enhanced elimination of the toxin

Specific therapy depends on the presumed toxin ingested

Mushrooms of Hallucinogenic Effect

The following mushrooms cause mild to fatal reactions and also hallucinogenic effect (Figure 14):

- ☆ *Agaricus* species: *A.pilatianaus, A. xanthodermus*
- ☆ *Amanita* species: *A.aspera, A. brunnescens, A.muscaria,A.phalloides.*
- ☆ *Boletus* species: *B.purpureus, B.rhodoxanthus, B.satanas, B.satanoides.*
- ☆ *Clitocybe* species: *C. acromelalgia, C. dealbata, C.rivulosa.*
- ☆ *Lepota* species: *L.brunneo-incarnata.*
- ☆ *Paxillus* species: *P.involutus.*
- ☆ *Psilocybe* species: *P. semilanceata, P. cubensis.*
- ☆ *Russula* species: *R. lividus.*
- ☆ *Tricholoma* species: *T. muscarium, T. pardinum.*
- ☆ *Volvariella* species: *V.media, V. parvula.*

When anyone, after eating mushroom develops symptoms, no time should be lost in calling a doctor. In the meantime, preliminary treatment may be given to remove remains of eaten mushroom.

Mushrooms which Cause Great Discomfort

The following species may cause great discomfort, sometimes requiring hospitalization, but are not considered deadly.

- ☆ *Amanita muscaria* (fly agaric) contains the psychoactive muscimol and the neurotoxin ibotenic acid. Ibotenic acid decarboxylates into muscimol upon curing of the mushroom, rendering it relatively non-toxic, though death via respiratory depression is possible. Muscimol intoxication is often considered unpleasant and undesirable however, and as such has seen little recreational use compared to the unrelated psilocybin mushroom, though it has been used as an entheogen by the native people of Siberia.

- ☆ *Amanita pantherina* (panther mushroom) contains similar toxins as *A. muscaria*, but is associated with more fatalities than *A. muscaria*.

  http://en.wikipedia.org/wiki/Mushroompoisoning-cite note-emediaHallu-14.

- ☆ *Chlorophyllum molybdites* (greengills) causes intense gastrointestinal upset.

- ☆ *Entoloma* (pinkgills) – some species are highly poisonous, such as livid entoloma *(Entoloma sinuatum), Entoloma rhodopolium*, and *Entoloma nidorosum*. Symptoms of intense gastrointestinal upset appear after 20 minutes to 4 hours, caused by an unidentified gastrointestinal irritant.

  http://en.wikipedia.org/wiki/Mushroompoisoning-citenote-sydney-31.

Figure 14: Mushrooms Causing Mild to Fatal Reaction and Hallucinogenic Effect.

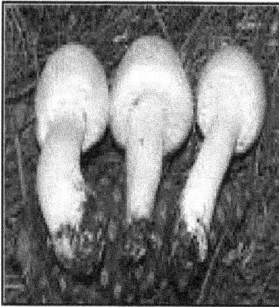

Agaricus xanthodermus

Agaricus pilatianaus

Amanita brunnescens

Amanita muscarisa

Amanita aspera

Amanita phalloides

Boletus purpureus

Boletus rhodoxanthus

Boletus satanas

Boletus satanoides

Clitocybe acromelalgia

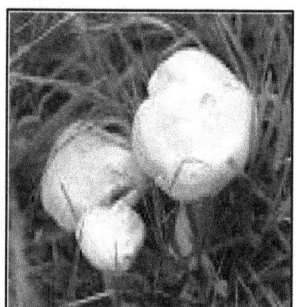

Clitocybe dealbata

Contd...

Figure 14–*Contd...*

Clitocybe rivulosa

Lepota brunneo-incarnata

Paxillus involutus

Psilocybe cubensis

Psilocybe semilanceata

Russula sp.

Tricholoma muscarium

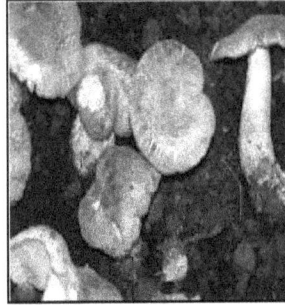

Tricholoma pardinum

Tricholoma sp.

Volvariella media

Volvariella parvula

☆ Many *Inocybe* species such as *Inocybe fastigiata* and *Inocybe geophylla* contain muscarine, while *Inocybe erubescens* is the only one known to have caused death.

☆ Some white *Clitocybe* species, including *C. rivulosa* and *C. dealbata* – contain muscarine.

http://en.wikipedia.org/wiki/Mushroompoisoning-citenote-Arora-10.

☆ *Tricholoma pardinum, Tricholoma tigrinum* (Tiger Tricholoma) causes gastrointestinal upset due to an unidentified toxin, begins in 15 minutes to 2 hours and lasts 4 to 6 days.

☆ *Tricholoma equestre* Man-on-horseback until recently thought edible and good, can lead to rhabdomyolysis after repeated consumption.

☆ *Hypholoma fasciculare/Naematoloma fasciculare* (Sulfur tuft) usually causes gastrointestinal upset but the toxins fasciculol E and F could lead to paralysis and death.

http://en.wikipedia.org/wiki/Mushroompoisoning- citenote-Arora-10.

☆ *Paxillus involutus* (Brown roll-rim) once thought edible, but now found to destroy red blood cells with regular or long-term consumption.

http://en.wikipedia.org/wiki/Mushroompoisoning-citenote- Bresinsky-19.

☆ *Boletus satanas* (Devil's bolete), *Boletus luridus, Boletus legaliae, Boletus piperatus, Boletus erythropus, Boletus pulcherrimus* causes gastrointestinal irritation. Of these, only *B. pulcherrimus* has been implicated in a death. Many books list *B. erythropus* as edible, but Arora lists it as "to be avoided."

http://en.wikipedia.org/wiki/Mushroompoisoning- citenote-Arora-10.

☆ *Hebeloma crustuliniforme* (known as Poison pie or Fairy cakes) causes gastrointestinal symptoms such as nausea and vomiting.

☆ *Russula emetica* (the Sickener) – as its name implies, causes rapid vomiting. Other Russulas with a peppery taste (*Russula silvicola, Russula mairei*) will likely do the same.

http://en.wikipedia.org/wiki/Mushroompoisoning-citenote-Arora-10.

☆ *Agaricus hondensis, Agaricus californicus, Agaricus praeclaresquamosus, Agaricus xanthodermus* – cause vomiting and diarrhea in most people, although some people seem to be immune.

http://en.wikipedia.org/wiki/Mushroompoisoning-citenote-Arora-10.

☆ *Lactarius piperatus, Lactarius torminosus, Lactarius rufus* – these and other peppery-tasting *Lactarius* are pickled and eaten in Scandinavia, but are indigestible or poisonous unless correctly prepared.

http://en.wikipedia.org/wiki/Mushroompoisoning-citenote-Arora-10.

☆ *Lactarius vinaceorufescens, Lactarius uvidus* – reportedly poisonous. Arora reports that all yellow- or purple-staining *Lactarius* are "best avoided."

http://en.wikipedia.org/wiki/Mushroompoisoning-citenote-Arora-10.

☆ *Ramaria gelatinosa* – causes indigestion in many people, although some seem immune.

http://en.wikipedia.org/wiki/Mushroompoisoning-citenote-Arora-10.

☆ *Gomphus floccosus* (the scaly chanterelle) – causes gastric upset in many people, although some eat it without problems. G. floccosus is sometimes confused with the chanterelle.

http://en.wikipedia.org/wiki/Mushroompoisoning-citenote-Arora-10.

5.3. Other Factors that Render Mushroom Poisonous

Mushrooms may be rendered poisonous by insecticides or herbicides sprayed on lawns or reserves. At least one author recommends never picking them in non-natural landscapes for this reason.

Also, mushrooms are sometimes contaminated by concentrating pollutants, such as heavy metals or radioactive material.

Rotten mushrooms may cause food poisoning. Mushrooms which are mushy, bad-smelling, or moldy (even of a choice edible species) may be toxic due to bacterial decay or mold.

Many mushrooms are high in fiber. Excessive consumption of mushrooms may lead to indigestion, which may be diagnosed as mushroom "poisoning."

5.4. Mushroom Toxins

Poisonous mushrooms contain a variety of different toxins that can differ markedly in toxicity. Symptoms of mushroom poisoning may vary from gastric upset to life-threatening organ failure resulting in death. Serious symptoms do not always occur immediately after eating; often not until the toxin attacks the kidney or liver, sometimes days or weeks later.

The most common consequence of mushroom poisoning is simply gastrointestinal upset. Most "poisonous" mushrooms contain gastrointestinal irritants which cause vomiting and diarrhea (sometimes requiring hospitalization), but usually no long-term damage. However, there are a number of recognized mushroom toxins with specific, and sometimes deadly, effects:

☆ Alpha-amanitin (deadly: causes liver damage 1–3 days after ingestion) – principal toxin in genus *Amanita*.

☆ Phallotoxin (causes gastrointestinal upset) – also found in poisonous Amanitas

☆ Orellanine (deadly: causes kidney failure within 3 weeks after ingestion) – principal toxin in genus *Cortinarius*.

☆ Muscarine (sometimes deadly: can cause respiratory failure) – found in genus *Omphalotus*.

☆ Gyromitrin (deadly: causes neurotoxicity, gastrointestinal upset, and destruction of blood cells) – principal toxin in genus *Gyromitra*.

☆ Coprine (causes illness when consumed with alcohol) – principal toxin in genus *Coprinus*.

☆ Ibotenic acid (causes neurotoxicity) and muscimol (causes CNS depression and hallucinations) – principal toxins in *Amanita muscaria, A. pantherina, and A. gemmata.*

☆ Psilocybin and psilocin (causes CNS arousal and hallucinations) – principal 'toxins' in psilocybin mushrooms, many of which belong to the genus *Psilocybe.*

☆ Arabitol (causes gastrointestinal irritation in some people).

☆ Bolesatine a toxin found in *Boletus satanas*

☆ Ergotamine (deadly: affects the vascular system and can lead to loss of limbs and death): An alkaloid found in genus *Claviceps*.

The symptoms of mushroom poisoning vary depending on the toxins involved.

☆ Alpha-amanitin: For 6–12 hours, there are no symptoms. This is followed by a period of gastrointestinal upset (vomiting and profuse watery diarrhea). This stage is caused primarily by the phallotoxins and typically lasts 24 hours. At the end of this, second stage is when severe liver damage begins. The damage may continue for another 2–3 days. Kidney damage can also occur. Some patients will require a liver transplant. *Amatoxins* are found in some mushrooms in the genus *Amanita*, but are also found in some species of *Galerina* and *Lepiota*. Overall, mortality is between 10 and 15 percent. Recently, Silybum marianum or blessed milk thistle has been shown to protect the liver from amanita toxins and promote regrowth of damaged cells.

http://en.wikipedia.org/wiki/Mushroompoisoning-citenote-benjamin95-23.

☆ Orellanine: This toxin causes no symptoms for 3–20 days after ingestion. Typically around day 11, the process of kidney failure begins, and is usually symptomatic by day 20. These symptoms can include pain in the area of the kidneys, thirst, vomiting, headache, and fatigue. A few species in the very large genus *Cortinarius* contain this toxin. People who have eaten mushrooms containing *orellanine* may experience early symptoms as well, because the mushrooms often contain other toxins in addition to *orellanine*.

http://en.wikipedia.org/wiki/Mushroompoisoning-citenote-ClinTox2001-6.

http://en.wikipedia.org/wiki/Mushroompoisoning-citenote-emedOrell-26.

A related toxin that causes similar symptoms but within 3–6 days has been isolated from *Amanita smithiana* and some other related toxic *Amanitas*.

http://en.wikipedia.org/wiki/Mushroompoisoning-citenote-pelzzaria-27.

☆ Muscarine: Muscarine stimulates the muscarinic receptors of the nerves and muscles. Symptoms include sweating, salivation, tears, blurred vision, palpitations, and, in high doses, respiratory failure.

http://en.wikipedia.org/wiki/Mushroompoisoning-citenote-emed-28.

Muscarine is found in mushrooms of the genus *Omphalotus*, notably the Jack o' Lantern mushrooms. It is also found in *A. muscaria*, although it is now known that the main effect of this mushroom is caused by ibotenic acid. Muscarine can also be found in some *Inocybe* species and *Clitocybe* species, particularly *Clitocybe dealbata*, and some red-pored *Boletes*.

http://en.wikipedia.org/wiki/Mushroompoisoning-citenote-Arora-10.

☆ Gyromitrin: Stomach acids convert gyromitrin to monomethylhydrazine (MMH), a compound employed in rocket fuel. It affects multiple body systems. It blocks the important neurotransmitter GABA, leading to stupor, delirium, muscle cramps, loss of coordination, tremors, and/or seizures. It causes severe gastrointestinal irritation, leading to vomiting and diarrhea. In some cases, liver failure has been reported. It can also cause red blood cells to break down, leading to jaundice, kidney failure, and signs of anemia. It is found in mushrooms of the genus *Gyromitra*. A gyromitrin-like compound has also been identified in mushrooms of the genus *Verpa*.

http://en.wikipedia.org/wiki/Mushroompoisoning-citenote-FDAimport-15.

☆ Coprine: Coprine is metabolized to a chemical that resembles disulfiram. It inhibits aldehyde dehydrogenase (ALDH), which generally causes no harm, unless the person has alcohol in their bloodstream while ALDH is inhibited. This can happen if alcohol is ingested shortly before or up to a few days after eating the mushrooms. In that case the alcohol cannot be completely metabolized, and the person will experience flushed skin, vomiting, headache, dizziness, weakness, apprehension, confusion, palpitations, and sometimes trouble breathing. Coprine is found mainly in mushrooms of the genus *Coprinus*, although similar effects have been noted after ingestion of *Clitocybe clavipes*.

☆ Ibotenic acid: Decarboxylates into muscimol upon ingestion. The effects of muscimol vary, but nausea and vomiting are common. Confusion, euphoria, or sleepiness are possible. Loss of muscular coordination, sweating, and chills are likely. Some people experience visual distortions, a feeling of strength, or delusions. Symptoms normally appear after 30 minutes to 2 hours and last for several hours. *A. muscaria*, the "Alice in Wonderland" mushroom, is known for the hallucinatory experiences caused by muscimol, but A. *pantherina* and *A. gemmata* also contain the same compound. While normally self-limiting, fatalities have been associated with *A. pantherina*, and consumption of a large number of any of these mushrooms is likely to be dangerous.

http://en.wikipedia.org/wiki/Mushroompoisoning-citenote-Arora-10.

http://en.wikipedia.org/wiki/Mushroompoisoning-citenote-emedHallu-14.

☆ Psilocybin: Dephosphorylates into the psychoactive psilocin upon ingestion, which acts as a psychedelic drug. Symptoms begin shortly after ingestion. The effects can include euphoria, visual and religious hallucinations, and heightened perception. However, some persons experience fear, agitation, confusion, and schizophrenia-like symptoms. All symptoms generally pass after several hours. Some (though not all) members of the genus *Psilocybe* contain psilocybin, as do some *Panaeolus*, *Copelandia*, *Conocybe*, *Gymnopilus*, and others. Some of these mushrooms also contain baeocystin, which has effects similar to psilocin.

☆ Arabitol: A sugar alcohol, similar to mannitol, which causes no harm in most people but causes gastrointestinal irritation in some. It is found in small amounts in oyster mushrooms, and considerable amounts in Suillus species and Hygrophoropsis aurantiaca (the "false chanterelle").

http://en.wikipedia.org/wiki/Mushroompoisoning-citenote-SporePrint-29.

Some mushrooms contain less toxic compounds and, therefore, are not severely poisonous. Poisonings by these mushrooms may respond well to treatment. However, certain types of mushrooms, such as the Amanitas, contain very potent toxins and are very poisonous; so even if symptoms are treated promptly mortality is high. With some toxins, death can occur in a week or a few days. Although a liver or kidney transplant may save some patients with complete organ failure, in many cases there are no organs available. Patients who are hospitalized and given aggressive support therapy almost immediately after ingestion of amanitin-containing mushrooms have a mortality rate of only 10 per cent, whereas those admitted 60 or more hours after ingestion have a 50–90 per cent mortality rate.

http://en.wikipedia.org/wiki/Mushroompoisoning-citenote-FDAbadbug-30.

5.5. Symptomology Based Antidotes (Supportive Measures)

1. **Methemoglobinemia:** which may occur after the ingestion of gyromitrins and occasionally after an intravenous (IV) spread of psilocybin, is treated with IV methylene blue. The US Food and Drug Administration (FDA) warns against the concurrent use of methylene blue with serotonergic psychiatric drugs, unless such therapy is indicated for life-threatening or urgent conditions. Methylene blue may increase CNS serotonin levels, increasing the risk of serotonin syndrome.

2. **Hemolysis:** which may occur with gyromitrin toxicity, is usually mild, necessitates the administration of large amounts of IV fluids only to prevent renal complications; blood transfusions are rarely required. Hemolysis due to *Paxillus* species may be more severe and may result in acute renal failure.

3. **Rhabdomyolysis:** has been reported with several species. Direct damage to myocytes with resultant onset on rhabdomyolysis occurs after ingestion of the so-called "man-on-horseback" mushroom, *Tricholoma equestre* (also

known as*Tricholoma flavovirens*). Patients may present with muscle pain and have been reported with elevated creatinine phosphokinase levels, (10,000 U/L to 100,000 U/L range). Other mushrooms implicated in less severe forms of rhabdomyolysis are *Russula subnigricans* (blackening Russula), *Boletus edulis* (king boletus), *Leccinium versipelle* (brown birch boletus), and *Albatrellus ovinus*(sheep polypore). Many of these are identified in field guides as edible. Treatment is with aggressive IV fluid resuscitation and consideration for IV sodium bicarbonate to alkalinize the urine. In rare cases, dialysis may be needed if renal failure occurs.

4. **Agitation:** commonly observed with hallucinogenic mushrooms, is treated with benzodiazepines; phenothiazines are best avoided in this setting. Other causes of agitation (eg, hypoxia, hypovolemia, and shock) should also be sought and corrected.

5. **Anticholinergic:** poisoning may be treated with benzodiazepines; in rare cases, physostigmine may be required.

6. **Severe muscarinic symptoms:** may be treated with the infusion of small doses of atropine. In muscarine poisoning, the entire episode usually subsides in 6-8 hours; some symptoms may take up 24 hours to fully resolve. Atropine should be considered only when excessive bronchial secretions compromise breathing and cause shortness of breath. Monitoring with pulse oximetry is indicated. Clinicians should be prepared to support the airway and perform orotracheal suctioning if necessary.

7. **Patients with severe poisoning from disulfiram-containing mushrooms:** may benefit from fomepizole (4-methylpyrazole), which blocks alcohol dehydrogenase and, hence, the formation of the toxic aldehyde.

8. **Fulminant hepatic failure (FHF):** is a common complication observed with amatoxin and gyromitrin poisoning, and it should be treated aggressively because it commonly follows a fatal course. Orthotopic liver transplantation (OLT) may be indicated.

9. **Renal failure:** commonly observed with norleucine and orellanine poisoning, may have to be treated with hemodialysis. Acute renal failure (ARF) with mild reversible liver injury may also follow the ingestion of *Amanita smithiana* and *Amanita proxima*.

   Conventional indications for dialysis include uremic encephalopathy, fluid overload (with pulmonary edema), severe hyperkalemia, and acidosis. Patients with unremitting renal failure are candidates for renal transplantation, but since most cases resolve slowly over time, several months of hemodialysis should occur before this is considered.

   The development of renal failure in patients with FHF warrants an attentive search for the cause of the renal failure. Patients with hepatorenal syndrome (HRS) are candidates for liver transplant.

Intensive Care in Mushroom Poisoning

Endotracheal intubation is recommended in all patients at risk of aspiration, and mechanical ventilation should be initiated in all patients with hypoxia, hypercarbia, acidemia, and shock. Aggressive rehydration in the intensive care unit (ICU) may be necessary in patients with cholera like gastroenteritis, and infusions of large amounts of electrolytes with dextrose solutions may be necessary to maintain vital functions.

Blood transfusions may be required in patients with hemorrhagic diarrhea, blood loss, and severe hemolytic anemia. Blood pressure support with dopamine and norepinephrine may be required when crystalloids and colloid infusions fail. Hypoglycemia is treated with infusions of 10 per cent dextrose.

Cerebral edema is also treated in a conventional manner, which is aimed at reducing intracerebral pressure and preventing herniation. Hyperventilation, fluid restriction, osmotic diuresis, hypertonic saline, positioning the head of the bed at 30° from the horizontal plane, barbiturate coma, and anticonvulsants may be necessary.

Montanini *et al.* (1999) reported use of acetylcysteine as the life saving antitode in Amanita phalloides (death cap) poisoning caused by alpha-Amanitin an amatoxin known to produce deleterious effects on the liver and the kidneys, when circulating in the blood. Therapeutic options employed to treat mushroom intoxication, such as haemodiaperfusion on activated charcoal, high dosages of penicillin G, oral charcoal, etc., very often failed to act properly and liver transplantation (when a graft is available) appeared to be the only solution. In recent years, it has been postulated that the oxidant effects of alpha-amanitin could be counteracted by the use of antioxidants such as silibinin. In a clinical trial by S. Todd Mitchell of Dominican Hospital in Santa Cruz, Calif., all of the 60-some patients who took a compound called silibinin within four days of eating a death cap mushroom survived. The drug works by preventing the mushroom's toxins (amatoxins) from getting into the liver. Amatoxins themselves prevent liver cells from creating a necessary enzyme. The amatoxins then leech into bile, ending up back in the intestines, creating a cycle of poisoning. The toxins also progressively damage the kidneys, unless the victim drinks large amounts of water—or "aggressively hydrates". High dosages of N-acetyl-cysteine (CAS 616-91-1, NAC), (fluimucil) already used as antioxidant in paracetamol poisoning, were successfully used in Intensive Care Unit (ICU)(Montanini et. al. 1999) in the treatment of *Amanita phalloides* poisoning with a protocol (haemodiaperfusion on activated charcoal, high dosages of penicillin G, etc.)

5.6. Historical Mushroom Poisoning

☆ Siddhartha Gautama (known as The Buddha), by some accounts, may have died of mushroom poisoning around ~479 BCE, though this claim has not been universally accepted.

http://en.wikipedia.org/wiki/Mushroompoisoning-citenote-Stamets-33.

http://en.wikipedia.org/wiki/Mushroompoisoning- citenote-Wasson-34.

☆ Roman Emperor Claudius is said to have been murdered by being fed the death cap mushroom. However, this story first appeared some two centuries after the events, and it is even debatable whether Claudius was murdered at all. Pope Clement VII is also rumored to have been murdered this way. However, it is similarly debated whether he died from any kind of poisoning at all.

http://en.wikipedia.org/wiki/Mushroompoisoning-citenote-35.

☆ Holy Roman Emperor Charles VI and Tsaritsa Natalia Naryshkina are believed to have died from eating the death cap mushroom.

☆ According to a popular legend, the composer Johann Schobert died in Paris, along with his wife, one of his children, maid servant and four acquaintances after insisting that certain poisonous mushrooms were edible.

☆ The best-selling author Nicholas Evans (The Horse Whisperer) was poisoned after eating *Cortinarius speciosissimus*. The parents of the physicist Daniel Gabriel Fahrenheit, who created the Fahrenheit temperature scale, died in Danzig on 14 August 1701 from accidentally eating poisonous mushrooms.

6

Mushroom: Global and National Scenario

6.1 Global Scenario

The global mushroom production as per FAO Statistics was estimated at about 2.18 to 3.41 million tons over period of last ten years (1997-2007). Current production too be around 3.5 million tons. China, USA, Netherlands, Poland, Spain, France, Italy, Ireland, Canada and UK are the leading producers (Table 4).

The three major mushroom producing countries as per FAO data *viz.*, China, USA, and Netherlands account for more than 60 per cent of the world production; however, share of China itself is 46 per cent which is about half of the world production. According to current Indian estimates, mushroom production of India is about 1 lakh metric tons, which is about 3 per cent of the world mushroom production.

In USA and Europe major contribution towards mushroom production is by white button mushroom. In Asian countries the scenario is different and other species are also cultivated for commercial production.

World mushroom production (FAOStat) is continuously increasing from 0.30 to 3.41 million tons over period of about last 50 years from 1961 to 2010. Also the export and import trend shows that the mushroom export/import has continuously increased in last 40 years, but marginally upto 1985 and beyond it there is tremendous increase in mushroom export/import upto 2010. Poland, Netherland, Ireland, China, Belgium, Lithuania, Canada and USA are the major mushroom exporting countries while countries like UK, Germany, France, Netherland, Belgium, Russian Federation and Japan import the mushroom from above exporting countries.

Table 4: World Production of Mushrooms (metric tons)

Countries	1997	2007
China	5,62,194	15,68,523
United States of America	3,66,810	3,59,630
Netherland	2,40,000	2,40,000
Poland	1,00,000	1,60,000
Spain	81,304	1,40,000
France	1,73,000	1,25,000
Italy	57,646	85,000
Ireland	57,800	75,000
Canada	68,020	73,257
United Kingdom	1,07,359	72,000
Japan	74,782	67,000
Germany	60,000	55,000
Indonesia	119,000	48,247
India	9,000	48,000
Belgium	NA	43,000
Australia	35,485	42,739
Korea	13,181	28,764
Iran	10,000	28,000
Hungry	13,559	21,200
Viet Nam	10,000	18,000
Denmark	8,766	11,000
Thailand	9,000	10,000
Israel	1,260	9,500
South Africa	7,406	9,395
New Zealand	7,500	8,500
Switzerland	7,239	7,440
Other countries	85911	59297
Total World Production	**21,86,222**	**34,14,392**

Source: World mushroom and truffles: Production, 1961- 2007;United Nation, FAO, FAOStat (8/28/2009).

World processed (canned and dried) mushroom is continuously increasing from 0.049 to 0.683 million tons over the period of last four decades (1970-2010) as compared to the fresh mushroom export (0.014 to 0.482 million tons) but fluctuation in export is higher in case of the processed mushroom. In USA, five decades ago, 75 per cent of the mushroom consumption was in the form of canned mushroom. Today, canned mushroom contributes only 15 per cent of total mushroom consumption. The consumption of canned mushroom is static and that of fresh mushroom has increased continuously. This clearly shows that consumer's interest is shifting towards fresh mushroom consumption.

Largest importer of preserved mushrooms (canned) is Germany with about 1,05,186 tons in 2007 (FAO) followed by Russian Federation (69,726 metric tons), USA (67,058 metric tons) and Japan (32,757 metric tons). Most of these supplies are made by China (4,05,112 metric tons), Netherlands (1,15,349 metric tons), Spain (20,623 metric tons), France (18,495 metric tons) and Indonesia (18,392 metric tons).

China is the largest producer and consumer of mushrooms in the world (15,68,523 metric tons production +17,732 metric tons imports) followed by USA (3,59,630 metric tons production +68,123 metric tons imports) and Netherland (2,40,000 metric tons production +7,884 metric tons imports) respectively.

Trade of mushrooms in European Union is significant (Table 5). The European Union mushroom production is about 27 per cent of the world production (2007, FAO). Netherlands is the largest producer and consumer, Poland is largest exporter, UK largest importer, France and Spain are also the larger producers as well as consumers. From outside, China is largest exporter of processed mushroom. Per capita consumption is very high (about 3.5 kg) in these countries (Table 6). Highest per capita consumption of mushroom is in Netherland (11.62 kg) followed by Ireland (6.10 kg) and Belgium (4.46 kg). Per capita consumption of mushroom in India has increased from 25 g to 40 g in last 10 years (1996-2007). However, as per Indian estimates per capita consumption in India is about 90g, which is very less compared to other countries including USA (1.49 kg) and China (1.16 kg).

Table 5: Mushrooms Trade in European Union (2007)

Country	Production (metric tons)	Export (metric tons)	Import (metric tons)	Consumption (metric tons)
UK	72,000	384	99,552	1,71,168
Belgium	43,000	36,918	33,876	39,958
Netherlands	2,40,000	83,592	34,161	1,90,569
Germany	55,000	4,553	54,168	1,04,615
Italy	85,000	3,135	9,835	92,600
France	1,25,000	3,013	37,158	1,59,145
Spain	1,40,000	848	1,688	1,40,840
Poland	1,60,000	1,47,817	1,213	13,396

It is clear from the Table 5 that European Union and USA are the biggest markets and Poland and China are the biggest competitors, for mushroom from India.

6.2 National Scenario

The production of mushroom, especially of the white button mushroom, in India has gone up in the last few years. There have been frequent reports of gluts in north Indian States during the winter months forcing the distress sale of the mushrooms. Therefore increase in the production without solving its market problems, would be counter-productive. The marketing of fresh mushrooms would determine the future of mushroom industry in India.

Table 6: Per Capita Consumption of Mushrooms (kg)

Country	1996	2007		
	Total	Fresh	Canned	Total
European Union				
UK	3.35	2.81	0.20	3.01
Belgium	3.90	3.71	0.69	4.46
Netherlands	2.80	11.62	0.00	11.62
Germany	3.83	1.27	1.20	2.47
Italy	2.60	1.56	0.06	1.62
France	3.03	2.58	0.14	2.72
Spain	–	3.11	0.00	3.11
Denmark	2.90	3.22	0.66	3.89
Ireland	–	6.05	0.05	6.10
Poland	–	0.35	0.00	0.35
USA	–	1.27	0.22	1.49
Canada	–	1.71	0.59	2.30
Japan	0.03	0.60	0.26	0.86
China		1.16	0.00	1.16
India		0.04	0.00	0.04

Table 7: State-wise Mushroom Production in India (tons) (2010)

Sl.No.	State	Button	Oyster	Milky	Other Mushroom	Total Production
1.	Andhra Pradesh	2,992	15	15	0	3,022
2.	Arunachal Pradesh	20	5	0	1	26
3.	Assam	20	100	5	0	125
4.	Bihar	400	80	0	0	480
5.	Chattisgarh	0	50	0	0	50
6.	Goa	500	20	0	0	520
7.	Gujarat	0	5	0	0	5
8.	Haryana	7,175	0	2	0	7,178
9.	Himachal Pradesh	5,864	110	0	2	5,993
10..	J&K	565	15	0	0	580
11.	Jharkhand	200	20	0	0	220
12.	Karnataka	0	15	0	0	25

Contd...

Table 7–*Contd...*

Sl.No.	State	Button	Oyster	Milky	Other Mushroom	Total Production
13.	Kerala	0	500	0	0	800
14.	Maharashtra	2,725	200	0	0	2,975
15.	Madhya Pradesh	10	5	0	0	15
16.	Manipur	0	10	50	50	60
17.	Meghalaya	25	2	0	0	27
18.	Mizoram	0	50	0	0	50
19.	Nagaland	0	75	250	250	325
20.	Orissa	36	810	5,000	5,000	5,846
21.	Punjab	58,000	2,000	0	0	60,000
22.	Rajasthan	100	10	10	10	120
23.	Sikkim	1	2	0	0	3
24.	Tamil Nadu	4,000	2,000	0	0	6,500
25.	Tripura	0	100	0	0	100
26.	Uttarakhand	8,000	0	0	0	8,000
27.	Uttar Pradesh	7,000	0	0	0	7,000
28.	West Bengal	50	50	0	0	100
	Union Territories					
1.	A&N Island	0	100	0	0	100
2.	Chandigarh	0	0	0	0	0
3.	Dadar and Nagar Haveli	0	0	0	0	0
4.	Daman and Diu	0	0	0	0	0
5.	Delhi	3,000	50	0	0	3,070
6.	Lakshadweep	0	0	0	0	0
7.	Puducherry	0	0	0	0	0
	Total	**1,00,683**	**6,399**	**920**	**5,313**	**1,13,315**

6.3 Production and Marketing

Fresh mushrooms have very short shelf-life, cannot be transported to long distance without refrigerated transport facility and are sold in localised markets in and around production areas. The cultivation of white button mushrooms throughout the year under controlled condition is restricted to few commercial units and 30-40 per cent of the production is being done under natural conditions during the winters. All the problems of marketing is experienced in 2-3 winter months (Dec-Feb) when more than 75 per cent of the annual production comes in market for sale in limited duration and market area. Farmers face the consequences of over-saturated market and are forced to sell their produce at highly unremunerative prices.

Market of mushrooms in India is not yet organized. It is the simple system of producers selling directly to retailer or even to consumer, which has its own limitations. Unlike the other countries where 10 per cent of the total cost is earmarked for marketing, we have not given marketing, sufficient thought and investment. Per capita consumption of mushrooms in India is less than 50g as against over a kg in various countries. There has not been any serious effort to promote the product, to strengthen and expand the market in order to increase its consumption. Mushroom is novel food item for this country and what to ask of its flavour, texture, nutritive value, many are not aware of 'what is mushroom and whether vegetarian or non-vegetarian item?'

In the coming years, there is going to be good demand for processed and fast foods. Mushrooms may be canned to meet the demand in the off-season and in the non-producing areas. Regarding the problems of sale/export of canned mushrooms, serious thought has to be given to bring down the cost of production of mushrooms and its processing in order to compete in the international market.

There is not much problem in the sale of fresh *Pleurotus* due to low production but there have been problems in selling dried 'Dhingri' particularly its export where middlemen take lion's share. Generally, the export orders are too big to be met by a single grower. *Pleurotus* growers may form a cooperative where they may pool their product and trade. APIDA and other Central as well as State agencies would be too willing to help them once they are assured of sufficient consignment for export, for 2-3 years.

7

Mushroom Farm Design

Mushroom cultivation is of recent origin in India. It is usually an indoor horticultural activity and a wide variety of designs and constructional materials have been used for mushroom houses. Construction of mushroom farm comprises various operations related to mushroom cultivation may be done efficiently to get mushroom production all the year around. Business point of view mushroom farm should not be constructed near the other mushroom farm as it adversely affects the consumption of the produce in the market. With the advent of modern cultivation technology, it is now possible to cultivate the mushroom throughout the year by employing environmentally controlled conditions. For obtaining profit, one start with at least three growing rooms and one-bulk chambers.

7.1. Selection of Site

When selecting a site to build a mushroom farm one has to keep in mind that following points are essential:

☆ Availability of good quality substrate
☆ Availability of clean water
☆ Availability of labour
☆ Adequate transport of the product to the market.
☆ Farm layout

Before planning the layout it is essential to decide whether compost preparation will take place at the farm. If this is the case, keep in mind that storage of basic materials as well as the composting site itself should be located as far away as possible from the growing rooms. It is equally important to know whether spawn will be purchased or prepared by the grower himself. In this case it is strongly advised that the spawn laboratory should not be located at the farm site at all, in order to prevent contamination spreading from one unit to another.

7.2. Design of Farm

The Dutch farm design and construction is being widely adopted all over the world with the modification to suit local conditions (Figures 15 and 16). There are two systems–single zone system and second zone system, which are generally used in mushroom growing farm.

7.2.1. Single Zone System

In this system, peak heating, spawn run and cropping is done in one room. Each room must be insulated properly and equipped so that the highest temperature required can be maintained. In this system, growing is always done in fixed beds called shelf beds.

7.2.2. Second Zone System

In this system, for peak heating, spawn run and cropping separate rooms are maintained. Temperature in cropping rooms will not rise than 16-20°C and heavy insulation is not required. To maintain desired temperature in peak heating and spawn run, the rooms are heavily insulated.

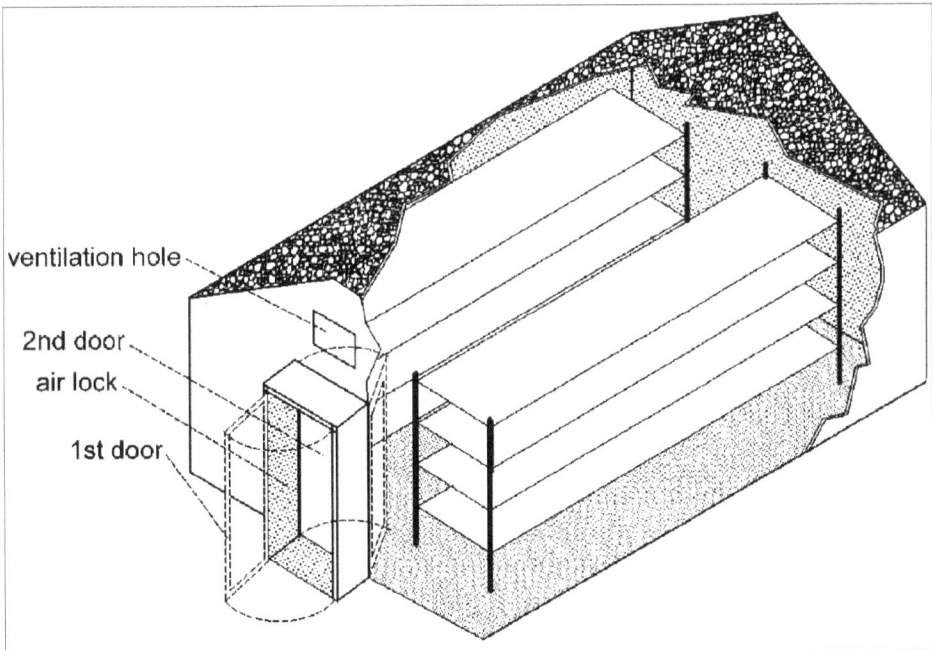

Figure 15: An Example of a Sophisticated Farmhouse Unit/Growing Room with Air Lock and Racks with Shelves.

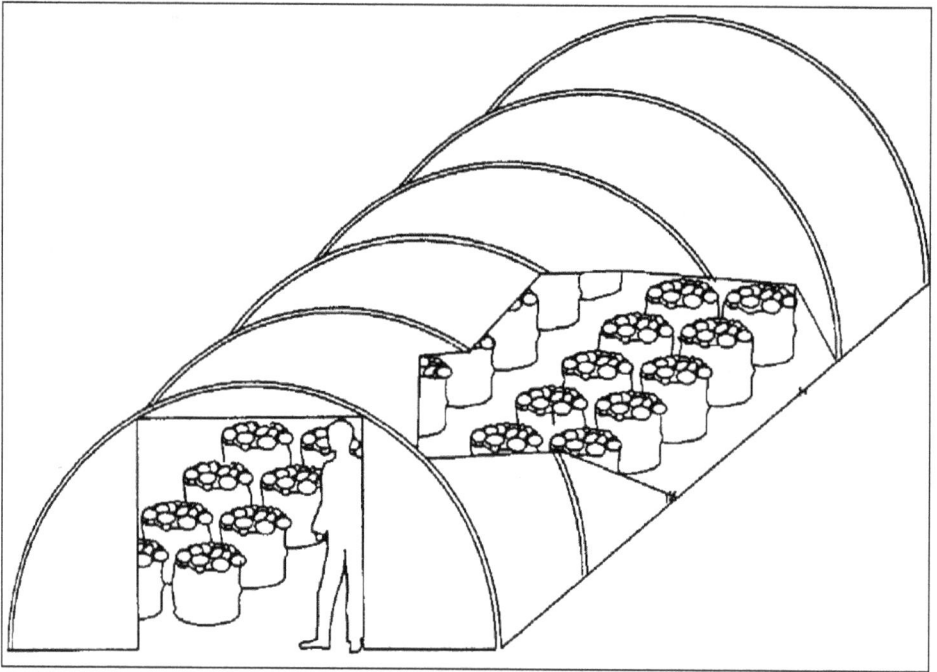

**Figure 16: An Example of a Low-Cost Mushroom Shed from Plastic
with Bags Placed on the Floor.**

7.2.3. Mushroom Farm Bulk Chamber

In a mushroom farm the bulk chamber is important, depends on the compost to be loaded. In Indian conditions, for 20 tons capacity, bulk chamber with the dimensions of 36′ (Long) ×9′ (Wide) × 12′ (Height) is best suited. The wall should be well insulated and the plenum should be deeper so that it give a desired slope for run off sufficient space of steam to penetrate the compost mass.

7.2.4. Temperature and Ventilation

Growing rooms at a mushroom farm should provide adequate climatic conditions. In particular ventilation and temperature are essential to ensure a reasonable production. In most western countries, mushroom growers make use of mechanical climate control but this requires high financial investments and therefore will not be treated in this book. To avoid high temperatures, more moderate temperature demanding mushrooms like *Agaricus* spp. are grown in caves or old tunnels or the farm can be built at higher and thus cooler altitudes. As most low-cost growing houses are constructed from bamboo, wood and plastic, a simple way of obtaining temperature reduction is by spreading wet sand on the floors underneath the shelves in the growing rooms and by wetting the bamboo-leave mats on the roof and the walls of the farmhouses.

7.2.5. Floors

Often, low cost growing houses are built just on bare ground. It is better to have a slightly tilted, cemented or concrete floor. These floors can be cleaned well and drain-water can flow out easily. Take good care that the drainage system of each room is not connected to another room, as diseases can spread easily through the draining pipe. For the same reason it is wise to frequently collect waste and contaminated material and to have them destroyed immediately after collecting.

7.3. Equipment and Facilities

During pasteurization, airflow of 150-200 m^3/hour per ton of fresh compost is required. Compost filling should be 900-1000 kgs/m^2 floor area with depth of 2.0-2.2m. In India, small chambers measuring 22'×8'×10' with iron door measuring 4'×6' has been used after providing proper insulation. The boiler of a usual capacity and coal/wood/oil fired will be enough. Oil fired to get temperature should be preferred. The blower has 24" fan with an opening fresh air from one side with provision to control air by adjusting shutter. The blower is run by 1440 rpm motor and pulling 6" ×3" is used to run blower.

7.4. Farm Hygiene

On a mushroom farm, hygiene is of vital importance. Since chemical control of pests and diseases is not feasible in small-scale mushroom cultivation, the only preventive measure is hygiene, and to some extent disinfection. This goes for a spawn production unit, the site for substrate production, the incubation rooms as well as for the production units.

7.5. Farm Location

Therefore checking a suitable site for a mushroom farm is very important. The surroundings of a farm should be clean and free from possible contamination from insects, moulds etc. This means that building a new farm close to other mushroom farms should be avoided. Insects and diseases from these farms could easily spread to the new farm. If possible separate the various operation units of the farm.

The spawn laboratory should be separate from the growing site. The growing rooms ought to be separated by closed (plastic) walls to keep the different stages of cultivation apart. As a matter of fact no incubation or spawn running should take place in the same room where the mushrooms are harvested. Debris, contaminated bags and spent mushroom substrate must be removed immediately from the rooms and from the farm itself, preferably to a place far away.

All these measures are necessary to avoid pests such as flies and mites as well as diseases spreading from these waste dumps. If the spent mushroom substrate is to be used for gardening soil, it should be transported as soon as possible and not be stored at the mushroom farm.

8

Spawn Production Technology

The spawn comprises mycelium of the mushroom and a supporting medium, which provides nutrition to the fungus during its growth. Modern mushroom production involves two cultivation steps, *viz.*, spawn making (which is the aseptic culture of mushroom mycelium) and mushroom growing (which is the growth and development of mycelium under protected but non- aseptic conditions to produce the fruiting bodies or mushrooms).

The spawn is mushroom seed and in mushroom industry it means the planting material, which consists of the vegetative stage of mushroom and its substrate. Duggar in 1905, made the significant discovery that a piece of the inner growing tissue of mushroom is capable of producing mycelium. The success of mushroom cultivation and its yield depend to a large extent on the purity and quality of the spawn used. Earlier virgin/natural spawn, brick spawn, flake spawn and mature spawn were used. Sinden in 1936, made use of cereal grains as the substrate for mushroom spawn production, he obtained patent for this and this revolutionized the mushroom industry. Now a day highly sophisticated, specialized and mechanized commercial spawn making system has been developed; and online global marketing is done. Before the introduction of cereal grain spawn, the spawn was prepared on sterilized composted manure and was known as manure spawn. Spawn production in bulk quantity for commercial use is much more laborious than for experimental purpose in smaller quantity. Strict hygiene must be observed when spawn has to be produced daily and also great care should be taken in the selection, raising and maintenance of cultures of mushroom strains otherwise there are chances of failure. The spawn production technology is divided into following steps.

8.1. Preparation of Grain Spawns

a) Raising and maintenance of pure culture

b) Master spawn preparation

c) Commercial spawn production

The preliminary requirement for making spawn is to obtain a pure culture of the desired mushroom species on a suitable medium, which is used for spawn preparation.

8.2. Pure Culture

Pure culture of mushroom mycelium can be raised either by tissue culture or spore culture raised from single or multispores collected from healthy and of desired characters mushroom. Multi-spore culture is obtained by placing a fresh fruit body after surface sterilization on a petriplate. Millions of spores are collected within 48 hrs. Serially diluted spores are then transferred to sterile Potato dextrose agar (PDA) or Malt-extract-agar culture slants. These slants are then incubated at $25\pm 1°C$ for two weeks to obtain pure culture. For tissue culture, the basidiocarp after surface sterilization is cut longitudinally and bits from the junction of stem and pileus are transferred to sterile Potato dextrose agar (PDA) or Malt-extract-agar (MEA) culture medium. The petriplates are incubated at $25\pm 1°C$ for two weeks. Mycelium from growing edges is carefully transferred to MEA/PDA slants and again incubated for 15 days to obtain pure cultures.

The growth of mycelium should be healthy, strandy, silky and off white to white in colour in case of *Agaricus bisporus*. The commercial spawn producer can also obtain the pure culture of the desired mushroom species from any established laboratory. Cultures are retrieved periodically for maintaining vigour and growth. For long time storage of cultures, these are preserved by specific methods.

8.3. Compositions of different Media

1. Malt Extract Agar (MEA)

☆ Malt extract - 25g

☆ Agar-agar powder - 20g

☆ Water -1 litre

2. Wheat Extract Agar (WEA)

☆ Wheat grains - 32g

☆ Agar-agar powder - 20g

☆ Water - 1 litre

Wheat grains are boiled in water for 1/2 hours and are filtered through the muslin or cheese cloth and wheat grains discarded. The volume of grain extract is raised to 1 litre with water and then agar-agar powder is added by stirring continuously. The medium is boiled for 5-10 minutes and pH is adjusted to 6.8-7.0.

3. Potato Dextrose Agar (PDA)
- ☆ Peeled Potato : 200g
- ☆ Dextrose : 20g
- ☆ Agar-agar Powder : 20g

Potatoes are peeled, washed, cut into small pieces (3-4c m) and boiled in water for 20 minutes. The potato extract is filtered through a cheese cloth and potato cubes discarded. The volume of this extract is raised to one litre. Dextrose and agar – agar powder is then mixed by stirring. The medium is boiled for 5-10 minutes and pH is adjusted to 6.8-7.0. Green potatoes should not be used as they contain anti-fungal alkaloids, which may be harmful to mushroom mycelium.

4. Compost Extract Agar
- ☆ Pasteurized and conditioned compost: 150 g (fresh weight)
- ☆ Agar-agar powder : 20 g
- ☆ Water : 1 litre

Pasteurized compost is boiled in one litre of water until the volume of water is reduced to half. It is filtered through cheese cloth and the volume of the compost extract is raised to one litre. Agar – agar powder is mixed by stirring and pH adjusted. Streptomycin sulfate @ 50mg/litre may be added after autoclaving to eliminate bacterial contamination.

5. Oat Meal Agar
- ☆ Oat meal flakes : 30g
- ☆ Agar – agar : 20g
- ☆ Water : 1 litre

8.4. Substrates for Spawn Production

Mushroom spawn can be prepared on any kind of cereal grains like wheat, *bajra* (pearl millet), *jowar* (sorghum), or *rye* etc. Though cereal grains are suitable for making spawn of any mushroom variety but due to cost, a variety of agricultural wastes like corn cobs, wooden sticks, rice straw, sawdust and used tea leaves etc. can also been used and variety of substrates for different mushrooms have been suggested. Sawdust of red tree wood species is preferred for making spawn.

8.5. Mother Spawn/Master Spawn Production

Well cleaned wheat grains are washed in clean, preferably chlorinated water and boiled in water (1:1.5w/v) for 15 minutes, left for another 10-20 minutes in the same hot water without boiling and then excess water is drained out by putting the contents on a wire mesh. This is done for having moisture content of grains around 50 percent and to make them soft for mycelial growth. After surface drying and cooling of the boiled grains, 2 per cent commercial grade gypsum and 0.5 per cent calcium carbonate (on dry wt basis) is added to these grains and thoroughly mixed. This will help to maintain the pH of the grain substrate and also to prevent sticking and clustering of grains. The substrate is now filled in wide mouthed glucose/milk bottles and plugged with non-absorbent cotton. These are then autoclaved at 22 lbs p.s.i. pressure for two hours in an autoclave, which will kill the contaminants present in the substrate. Some times endospores of *Bacillus* sp. survive at this pressure, so alternate sterilization (1h at 20 lbs psi) for two consecutive days requires to eliminate the *Bacillus* sp. completely. Autoclaved material is then taken out and then put in the inoculation room at room temperature for cooling till the substrate temperature is less than 25°C. After cooling bottles are inoculated with culture bits from pure culture of mushroom mycelium and incubated at 25°C. During incubation these bottles are shaken at weekly interval for complete colonization of grains by the mycelium. Thus master spawn is ready for use in 25-30 days.

8.6. Commercial Spawn

Commercial spawn can be prepared in heat resistant polypropylene bags. Normally the bags should be of 14×7 inch size or as per requirement. Polypropylene bags should have double sealing at the bottom and after filling of grains, they are plugged with the help of a polypropylene neck and non-absorbent cotton. The bags are then sterilized at 22 lbs p.s.i. pressure for 2 hours in an autoclave. Autoclaved bags are shaked well before inoculation so that the water droplets accumulated inside the bags are reabsorbed by the grains. The sterilized bags are kept in the laminar flow under UV light for about 20 minutes for surface sterilization. Fifteen to twenty grams of grains from master spawn bottles is inoculated per bag under aseptic conditions. One bottle of master spawn is sufficient for inoculating 25 to 30 commercial spawn bags. Inoculated bags are shaken well so that the inoculum is well mixed with other grains. However, to reduce the time period for spawn preparation, the quantity of inoculum may be increased. Then the bags are kept in incubation room for mycelium spread at 25°C. During incubation the bags are regularly examined for mould infestation and contaminated bags are immediately removed and autoclaved before discarding the bags to avoid build-up of contamination in the vicinity. Normally it takes 20-25 days for complete spread of mycelium of *Agaricus* and *Calocybe* species, whereas in case of *Pleurotus* spp. it takes about 10-15 days.

8.6.1. Precautions

1. Hygiene conditions should be maintained throughout the process of spawn preparation.
2. Selection of substrate should be done with utmost care (depending on the availability and cost). The grains must be bold. Broken, diseased, insect infested and shriveled grains should be avoided.
3. Washing of grains should be done with clean, preferably chlorinated water.
4. Packing material like polypropylene bags, neck rings, and non-absorbent cotton should be of good quality. Polypropylene bags must be checked for leakage before use.
5. Boiling of grains should be optimum and after boiling the grains should be surface dried and cooled before mixing gypsum and calcium carbonate.

6. Proper autoclaving is necessary. Do not release the pressure valve immediately after autoclaving, which causes water vapours to condense in the bottles.

7. Temperature of inoculation room should be kept low to avoid contamination.

8. The spawn laboratory should be dust free and fumigated with formaline regularly.

9. Bottles/bags should be never over sterilized as it causes loss of moisture in the grain which becomes very hard and such hard grains are not fully colonized by mycelium.

8.6.2. Characteristics of Good Spawn

1. There should be proper coating of mushroom mycelium around every grain used as a substrate for spawn production.

2. The growth of the mycelium in the spawn bottles/bags should be silky strand type. It should not be fluffy.

3. Spawn should be absolutely free from mould and bacterial growth. Fresh spawn has strong mushroom smell.

4. Spawn should not be more than one month old.

8.6.3. Facilities Required for Commercial Spawn Production

1. Washing and boiling room
2. Packing and Autoclaving room
3. Inoculation room
4. Incubation room
5. Cold store
6. Delivery room
7. Polypropylene bags/ bottles/glass wares
8. Steel racks
9. Chemicals
10. Non-absorbent cotton
11. Wheat grains

8.6.4. Instruments

1. Boiling kettles
2. LPG gas connection
3. Autoclave
4. Incubator
5. Refrigerator
6. Laminar flow
7. AC units

8. pH meter
9. Sieves

8.7. Transport of Spawn

During transportation spawn should not be exposed to temperature higher than 30°C. The spawn bottles or bags can be packed in thermocol boxes containing ice during summer months. Alternatively, spawn can be transported, during night. The spawn can be stored at 5°C for one month, if not used. *Volvariella* and *Calocybe* spawn should never be refrigerated. Once the container is opened, spawn should be used entirely.

8.8. Problems during Spawn Preparation

Microbial contamination is one of the main problem in the spawn lab. This may be due to bacteria or fungi. They causes 100 per cent losses as the contaminated spawn has to be discarded entirely. The common fungal contaminants during spawn preparation are *Aspergillus, Alternaria, Chaetomium, Cladosporium, Drechslera, Fusarium, Mucor, Rhizopus* and *Trichoderma*. This type of contamination can be recognized with the typical colours of their mycelium and spores. Such contaminated bags should be removed as and when they appear. This can be controlled by selection of good quality grains, proper autoclaving and hygienic conditions in lab. Bacterial contamination is more difficult to detect. If this is not detected in master bottles then all the commercial spawn prepared from these will be contaminated. *Bacillus* is common bacterial contamination and one can use ampicillin, streptocycline or streptomycine @50 mg/kg of boiled substrate before autoclaving. This also can be managed by maintaining pH of spawn substrate nearly 7.0 and incubating at 22-25°C.

When culture is maintained on same type of the medium for several generations, then it losses its metabolic potential. Therefore, it is advisable to change the culture medium or grow the culture on a compost based medium. The 10 days old culture tubes should be used for master spawn preparation.

8.9. Management of Contaminants

For the most common contaminants encountered during spawn preparation as enlisted above, grains are the main source of contamination. The fungal contamination can be seen with the typical colour of their mycelium, spores or conidia. If the contaminants are allowed to grow, they may spoil a large number of spawn bags. If such contaminated bags are not timely removed, it may become a perennial source of contamination. Selection of good quality grains, proper autoclaving and strict hygiene in spawn lab can reduce the contamination to a great extent. If the problem continues addition of carbendazim or thiophanate – methyl @ 0.15g/kg of boiled grains can help in reducing the losses caused by fungal contaminants.

Bacterial contamination is more difficult to detect. Some bacteria give a greasy appearance and emit a pungent or foul order smell. If the bacterial contamination is not detected in master spawn bottles, all the commercial spawn or compost prepared from them will become useless and may result into total loss of spawn or crop. Wet spot disease caused by *Bacillus species* is a common bacterial contamination problem in mushroom spawn. The disease can be managed by maintaining the pH of spawn substrate below 7.0 and at temperature 22-25°C. If the problem continues, antibacterial compounds like neomycin, streptomycin or streptocycline @10-50 µg/gram of spawn can be added after boiling of wheat grains to avoid bacterial contaminants.

8.10. Spawn Production as an Enterprise

1. Huge potential in spawn production.
2. More than 500 tons of the spawn required annually in Northern India alone.
3. Govt. is encouraging mushroom projects; thereby increasing demand of spawn.
4. Number of seasonal growers is increasing every year.
5. New agriculture policy stresses on diversification in agriculture in which mushroom cultivation occupies a prominent place.
6. Finance facilities available from nationalized banks.
7. Wide gap in demand and supply of mushroom at world level.
8. Scope of special mushrooms cultivation is still exploited in India.

9

Commercial Mushroom Production

9.1. Production of White Button Mushroom

The cultivation technique of white button mushroom involves four major components *viz.*, composting, spawning, casing and care during cropping. With a view to cut down the cost of cultivation of white button mushroom, which is much in demand, a low cost technology developed by CCS HAU can be used. This technology helps in bringing down the cost of cultivation substantially with more returns to the growers. Precisely, this low cost technology involves use of mustard straw (*Brassica*), for compost making, burnt rice husk for casing and thatched structure made up of locally available pearl millet, *jowar, sarkanda* or cotton sticks etc. for mushroom house.

White button mushroom requires an indoor temperature ranging between 15 to 25°C. It requires 22-25°C for vegetative growth and 14-18°C for fruiting. It can be grown conveniently during October to February under natural conditions in Northern India.

It is grown on the specially prepared substrate called compost. The compost is prepared by mixing different raw materials in specific proportion either by long or short method of composting. Since preparation of compost by short method requires specialized unit which is not feasible for small and marginal farmers, hence, only long method, which takes 28 days and involves 7-8 turnings at varying interval, is being used by seasonal mushroom growers.

9.1.1. Composting

Like other fungi, button mushroom (*Agaricus bisporus*) is heterotrophic organisms and requires readymade food for its growth. It requires carbon, phosphorus, sulphur, potassium, iron and vitamins such as thiamine for its growth. All the ingredients that contain these compounds are mixed in a fixed proportion and decomposed in a set pattern to form a substrate, which is selective in nature for supporting the growth of *Agaricus* mycelium. This substrate is called compost and process of decomposition in a set of pattern is called composting. The decomposition process is governed by a number of micro-organisms which produce important biochemical reaction thereby making it selective for *Agaricus* and exclusion of other competitive micro-organisms.

Initial attempt on commercial cultivation of *A.bisporus* was made on the horse manure. Horse stable manure has self heating properties which after decomposition become ideal for supporting the mycelia growth of *Agaricus* spp. The compost with horse stable manure as base material is termed as natural compost. In 1921 compost based on nitrogenous fertilizers and wheat straw was patented and termed as artificial or synthetic compost. This compost was further improved with the addition of the phosphate fertilizers and gypsum etc. Later on synthetic compost was prepared by various workers using different lignocellulosic agricultural wastes or by-product for making the compost. The composting process was also standardized and named as long method of composting (LMC), short method of composting (SMC) and indoor method of composting (IMC), based on time taken and technology employed. Now – a- days only synthetic compost made by any of the above method is used for the cultivation of button mushroom.

9.1.2. Materials Used for Composting

The ingredients used for making compost are broadly divided into base material, supplements and ingredients for rectifying mineral deficiency.

9.1.2.1. Base Materials

For the preparation of compost different agricultural by-products like cereal straw (wheat, barley, paddy), maize stalk, sugarcane baggage or any other cellulosic agro wastes can be employed. In North India mostly wheat, *Brassica* and paddy straw are used as base material. The straw to be used should not have been exposed to rains. It should have 5-8 cm long pieces, otherwise heap prepared by long straw would be leading to improper composting and if straw is too short, the heap will be too compact leading to anaerobic conditions. The base material provides a reservoir of cellulose, hemicelluloses and lignin, which are used by the mushroom mycelium as carbon source during its growth. These straws also provide bulk and proper physical structure to the substrate which facilitate aeration for the build up of desired microflora essential for composting. Paddy straw is very soft, degrade quickly during composting and also absorb more water as compared to wheat straw. While using the base materials care should be taken on the quantity of water to be used, schedule of turnings and adjustment to the rate and type of supplements.

9.1.2.2. Supplements

Since base materials do not have adequate amount of nitrogen and other components to start with the decomposition process, they are supplemented with nitrogen and carbohydrate rich materials to start the process. A large number of materials like animal manures are employed for this purpose.

9.1.2.2.1. Animal Manures

In general cattle dung is referred to as animal manure but it has not been found suitable for compost preparation. However, manure from horse, poultry, pig and sheep are considered as good supplements. Nitrogen content in these manures may vary from 1-5 per cent. Besides nitrogen they also provide little of carbohydrate and both of these are released slowly during the composting process. In addition of providing nutrients, they also enhance the bulk of compost which is very useful under our conditions when cost of wheat straw is increasing. In India, horse and poultry manure are mainly used as supplements. Horse manure along with bedding wetted with urine, is best suited for compost making and does not require any further supplementation. But when horse manure does not have enough bedding material, supplementation with inorganic nitrogen along with wheat straw may prove useful. Due to non-availability of good quality horse manure more and more mushroom growers are using poultry manure as a supplement. While selecting the poultry manure, care should be taken that it has more droppings than the litter. Poultry manure to be used must be got tested for nitrogen content which should be between 2.5-3.0 per cent and on the basis of test report the additional amount of nitrogen to be added should be computed.

Growers who have pasteurized tunnel facilities generally use poultry manure as supplement and harvest good yields. However, discourage the use of poultry manure in long method of composting, because poultry manure harbors many pathogenic nematodes and fungi, which ultimately reduce the mushroom yields. If it is to be used it must be treated with carbendazin and DDVP @ 0.05 and 0.1 percent before supplementation or steam pasteurized.

9.1.2.2.2. Carbohydrate Nutrients

Carbohydrates to the composting substrate may be supplied in the form of molasses (by product of sugar mills), wet brewer's grain (by product of distilleries), malt sprouts (by product of beer industry), wheat bran (by product of flour mills). These are essentially needed to correct C/N ratio and are also necessary for the establishment of bacterial flora in the compost. They also provide protein to the compost.

9.1.2.2.3. Nitrogenous Fertilizers

Nitrogen is essentially required for the microflora which carries out decomposition of the straw. For achieving quick establishment of desired microflora, sources which release nitrogen at a fast rate are required and for this, inorganic nitrogenous fertilizers are the best source. Therefore, fertilizers like ammonium sulphate, calcium ammonium nitrate and urea are generally used. Undoubtedly, N is the most important element of compost, but its excess use is still more harmful. In no case its percentage at the time of stacking should exceed 1.75 per cent (on dry wt. basis of base material) at start of compost preparation.

9.1.3. Materials to Correct Mineral Deficiencies

Besides carbon and nitrogen *A.bisporus* also require other elements like potash and phosphorus. These may be supplied to the compost in the form of murate of potash and super phosphate.

9.1.4. Materials to Correct Greasiness

Gypsum ($CaSO_4$): It's addition to the composting material provide trace elements (calcium) and also help in removing greasiness of compost by precipitating suspended colloidal materials.

9.1.5. Formulations

From time to time, different workers/research centers have suggested various compost formulations depending upon the availability of the basic materials in their respective areas. But mushroom growers are advised to select the basic material on the basis of availability and cost. The main objective of computing the formulation is to achieve a balance between carbon and nitrogen compounds. It has been found that N percentage of 1.5 to 1.75 at stacking gives the best result. Nitrogen level below 1.5 per cent gets easily contaminated with cellulose loving fungi particularly *Stachybotrys atra, Doratomyces stemonitis, Stilbum nanum* and *Trichoderma viride,* while N level above 1.75 per cent may make compost an easy source for *Sepedonium* spp. (yellow mould). Hence care should be taken to restrict the N level at the beginning of composting at 1.5-1.75 per cent.

Compost Formulations

1. *Natural compost*

Horse manure	1000kg
Wheat straw	350kg
Urea	3kg
Gypsum	30kg

2. *Synthetic compost*

i.

Wheat and Paddy straw (1:1)	300kg
CAN	9kg
Urea	3 kg
Wheat bran	30 kg
Murate of potash	3 kg
Gypsum	30 kg
Lindane dust (5 percent)	250 gm

ii.

Wheat straw	300kg
Poultry manure	60 kg
CAN	9 kg
Single super phosphate	3 kg
Wheat bran	30 kg
Gypsum	30 kg
Lindane dust (5 per cent)	250 gm

iii.

Wheat straw	300 kg
CAN	30 kg
Chicken manure	60 kg
Urea	5.5 kg
Gypsum	30 kg
Lindane dust (5 per cent)	250 gm

Formulae (CCS HAU)

i.

Wheat straw	300 kg
CAN	9 kg
Urea	3.5 kg
Single super phosphate	3 kg
Wheat bran	30 kg
Murate of potash	3 kg
Gypsum	25 kg
Molasses	5 kg

ii.

Wheat straw	300 kg
Chicken manure	60 kg
Wheat bran	7.5 kg
CAN	6 kg

Urea	2 kg
Murate of potash	2 kg
Gypsum	20 kg

iii.

Brassica straw	300 kg
Chicken manure	60 kg
Wheat bran	8 kg
Urea	4 kg
Single super phosphate	2 kg
Gypsum	20 kg
Molasses	5 kg

Formulae for Short Method of Composting (SMC)

I

Wheat straw	1000 kg
Chicken manure	100 kg
Brewer's grain	72 kg
Urea	14.2 kg
Urea	30 kg

II

Brassica straw	1000 kg
Chicken manure	400 kg
Wheat bran	70 kg
Urea	15 kg
Molasses	15 kg
Gypsum	30 kg

The straw should be preferably fresh, yellow and shining and if old straw is to be used it should not be exposed to the rain. Old straw which is no longer bright yellow and shining can be used only if it is tough. Brassica straw if used must be prepared by thresher and supplemented with chicken manure.

9.1.6. Purpose and Methods of Composting

Composting for mushroom cultivation (for white button mushroom) is done for three basic purposes:

1. It transforms the straw and manure into a substrate more suitable for the growth of *A.bisporus* than for many microorganisms whose presence in the substrate is unavoidable.

2. To create a favourable medium for desirable microbial flora which promotes the growth of *A.bisporus*. Protein in the form of countless dead bacteria (biomass) and other microorganisms is a vital source for mushroom nutrition.

3. Its composting temperature due to thermo-genic and chemical reaction is high enough to eliminate most harmful pests and diseases.

9.1.7. Long Method of Composting

9.1.7.1. Facilities Required

It is always better to make the compost on a cemented platform because it is easy to clean and carry out other operations (wetting and turning) of composting. If such platform is not available then brick layered platform can also serve the purpose. If either of these is not available, it does not mean that compost cannot be prepared. Any *kachha* platform can also be used after levelling, cleaning and spreading a plastic sheet on it. But the quality of compost prepared at *kachha* platform will not be as good as of cemented/brick layered platform. If the composting yard does not have roof, some provision to protect the compost during heavy rains should be made. Size of the platform may vary as per the requirement. Approximately one quintal of straw needs 20 sq.ft. of space for wetting. A large amount of water is also needed for wetting of straw, *e.g.* about five times of dry weight of straw (*i.e.* 500 lt of water for 100 kg of straw is required for proper wetting), however, paddy straw requires less water.

To give the compost pile a proper shape, wooden/iron mould is also required. The mould consists of three boards (two side boards and one end board). Side boards are placed 5' apart and held in position by 2 clamps, end board is placed on the smaller side or rectangular block. If mould is made of the iron sheet with angle iron support, clamp connection is not needed. While placing all the three boards these are slightly tilted towards inside.

9.1.7.2. Method

The first step in the compost preparation is to clean the composting yard properly and spray 2 per cent formalin 24 hours before use. On the following day straw is spread over the platform and wetting begins. Wheat and brassica straw being hardy in nature does not absorb water quickly. Therefore, frequent turning of straw with forks are must while sprinkling water. The wetting of straw should continue till it attains a moisture percentage of around 75 per cent. Wetting normally takes 24-48 hours. But, if even after 48 hours desired moisture content is not achieved wetting should be continued.

If the straw is not wetted properly in the beginning this will result in poor build up of microbial population and hence slower decomposing/composting. Excess water sprinkled after the start of decomposition process will wash down the nutrients, particularly nitrogen and useful microflora and ultimately deteriorate the quality of the compost. Therefore, light water spray on dry part of compost is recommended during periodic turnings.

Day-0

Although decomposition of straw starts when straw is wetted, yet scientifically on day -0 when wetted base material mixed with fertilizers and other supplements except gypsum are stacked in heap is considered as the starting point. Some growers prefer to mix half the supplements at the beginning of composting and the remaining half at/after the first turning.

Before making the heap all the ingredients added should be mixed thoroughly. It is always better to make the heap with the help of mould assembly but if moulds are not available the width and height of the heap should be kept about 4-5 feet. The length of the heap can be kept as per convenience. If composting is to be done in the cooler months/regions when temperatures ranges between 10-18°C, a small heap would be unable to retain heat and moisture and thus composting would be unsatisfactory. Generally during the hot weather and in tropical and sub-tropical regions in particular, the temperature difference between inside of the compost and surrounding air is too small to produce chimney-effect, necessary for compost ventilation. If core ventilation does not take place, undesirable acid zones will form inside the compost. In such case, relatively narrow heaps would be more suitable.

9.1.7.3. Turning Schedule

Soon after stacking of the compost on day -0, the compost temperature starts rising which after 24-48 hours reach around 70°C in the central portion. High temperature is necessary for proper composting; otherwise the compost will lack the necessary nutritional value so essential for a good crop. During composting process nitrogen in the form of ammonia is released which is utilized by the compost microflora and is assimilated in the form of protein (biomass) around straw. This protein is utilized by mushroom mycelium. The efficiency of compost microflora to utilize ammonia is maximum at 45-52 °C. To maintain this temperature in major portion of compost heap is given periodic turnings. Different workers from time to time have given different turning schedules. The schedule recommended by CCS HAU is as follow:

- ✰ Day-0–Wet mix and stack the heap
- ✰ Day-6–1st turning
- ✰ Day-10–2nd turning
- ✰ Day-13–3rd turning
- ✰ Day-16–4th turning
- ✰ Day-19–5th turning
- ✰ Day-22–6th turning

★ Day-25–7th turning

★ Day-28–filling of compost in trays/bags/beds

At third or fourth turning gypsum (powdered) is mixed in compost. During turning outer portion of the heap should be placed in the center of new pile. Every time while giving turning dry straw should be wetted. At the time of final turning compost should be free from the smell of ammonia and if the ammonia smell persists another turning at 1-2 days interval should be given.

9.1.8. Short Method of Composting

As the name indicates, in this the composting period is short. The method is divided into two phases: Phase-I and Phase-II.

The procedure for phase-I is similar to the initial stages of long method of composting except, that turnings are given sooner. Three turnings at an interval of three days are given; gypsum is added at third turning. The compost is now ready for phase –II. Previously, pasteurized rooms were in use whereas now a day's these have become obsolete. Bulk composting chamber or so called tunnels are in use these days. This method was developed in Italy and France and at present is successfully used in Netherlands. In this method the phase-II compost is loosely heaped on the grated floor of tunnel fixed on a plenum surface. Roughly 4 sq.ft. of tunnel surface can hold about one quintal of compost (dry wt. basis). Experiments conducted at CCS HAU, Hisar using low cost pasteurization tunnel revealed that after ten days of outdoor composting (phase-I) *i.e.* three turnings, thereafter the compost is filled in the tunnels up to the height of seven feet and the seven blower fan is put on for uniform circulation of air in the compost. The temperature of the compost is allowed to rise up to 56-58 °C and maintained for 4-6h, which is sufficient to kill pathogens and pests. Under our condition this temperature can be attained by the thermo-genic heat of compost and no boiler is required for giving steam. Later on temperature of the compost is maintained at 45°C for 6-7 days for conditioning of the compost or till ammonia smell disappears. Normally the temperature of the compost comes down as soon as the ammonia in the compost is utilized. Otherwise the compost temperature is lowered down by introduction of fresh air through the blower. The compost prepared by this method gives about double the yield of compost than prepared by long method.

9.1.9. Quality Assessment of Mushroom Compost

The good quality compost should have the following characteristics:

1. Physical

★ Colour - Dark brown

★ Smell - Pleasant

★ Pliability - Soft

★ Texture - Friable

2. Chemical

★ pH - 7.6-7.8

★ Ammonia - <0.006 per cent

☆ Nitrogen - 2.5 to 2.7 per cent

☆ Moisture - 68 to 70 per cent

3. Moisture

Moisture is an important factor in mushroom cultivation. The mushroom requires an atmosphere nearly saturated with moisture, yet the direct application of water on the beds is more or less injurious to the growing crop.

4. Ventilation

Ventilation is the most important factor governing crop production. Ventilation is responsible for making congenial environmental conditions and also for the removal of toxic gas by the introduction of adequate fresh air. The CO_2 level of 0.10 to 0.15 per cent volume is necessary during crop production and this can be achieved by giving 4 to 6 air changes /hour.

9.1.10. Spawning

Spawn (mushroom seed) is available on polypropylene bags. For 100 kg ready compost, 500g spawn is sufficient. After mixing spawn in the compost (when its temperature is around 25 °C), it is either filled in the polythene bags or it is spread on the racks. Then it is covered with old newspapers, which have been sprayed with 2 per cent formalin. The room temperature is maintained at 25 °C and humidity at 80-90 per cent by spraying water on the newspapers, walls and floor of the mushroom house. The height of the compost in the racks should be 6 inches, whereas in bags it should be 10-12 inches irrespective of the size of bag.

Compost preparation, in general should be started in 2nd week of September and spawning can be done in the second fortnight of October, as temperature is most suitable during this period in Northern India.

Method of Spawning

Method of spawning affects the yield of mushrooms:

1. **Double layer spawning:** In this spawning is done by scattering the spawn on racks beds when half filled with compost and then after the complete filling of the tray. The spawn is gently pressed uniformly each time and trays are covered with newspaper.

2. **Top spawning:** After filling the racks with compost, the spawn is planted just above the surface and then a thin but a better layer of compost is spread uniformly, because if the spawn is at the top of the compost it dries up quickly.

3. **Through spawning:** The grain spawn is mixed throughout the compost.

4. **Shake-up spawning:** Workers have reported a better productivity yield by shake-up spawning. After 7-10 days of spawning, the compost is thoroughly shaken up and after that either it is cased at once or a few days later.

5. **Active mycelium spawning:** In this method fully run trays/ bags of spawned compost are used for spawning new bags /beds. Thus one such

tray is used for several trays. In this method, however, chances of contamination are more.

6. **Spot spawning:** The grain spawn can be put in the holes at a certain distance with a pointed stick or fingers. The cavity is covered with the compost.

9.1.11. Casing

Casing means covering the compost with a thin layer of soil or soil like material after the spawn has spread in the compost. Casing is done for following purpose:

1. It gives support to the mushroom. Though mushroom can form in the uncased compost, it may fall due to its weight and supply of the food can be disturbed.

2. Casing soil provides humidity as it can hold water for a longer period.

3. It prevents quick drying of the spawned compost and therefore, it helps better spawn growth.

4. Vegetative mycelium is encouraged to fruit only when it enters into the medium which is deficient in nutrition. Casing soil provides such conditions.

5. Casing maintains the temperature. Reduction of temperature also encourages fruiting, soil losses moisture by evaporation and after each watering a cool layer is provide which appear to stock the warmth loving mycelium into activity.

Generally a mixture of FYM and garden soil/field soil is used. But a mixture of burn rice husk and garden soil/field soil (1:1) has been found to be cost effective. As burnt rice husk needs no sterilization, only garden soil is sterilized by formalin. After removing newspaper sheets or polythene sheets, a layer of 1-1.5" of this mixture is spread on compost. Casing should be done, when there is thorough/complete spawn run in the compost. After casing, water is to be sprayed on the racks/bags.

9.1.11.1. Sterilization of Casing Soil

To use it as a casing material it should be sterilized in such a way that harmful microorganisms are killed and the useful once remain. Sterilization of the casing material is done either by chemicals or by heating. Sterilization may also be done by steam from boiler through perforated pipes and temperature raised to 60°C and maintained for 5 hours. Sterilization under pressure is not advisable as the beneficial microorganisms are also killed and leave the soil more susceptible to re-infection.

Chemical Sterilization

Chemical commonly used for sterilization is formalin. For sterilization with formalin, about half a liter of formalin is diluted with 10 liters of water and used for 1 cubic meter of casing soil. The casing soil is spread over a plastic sheet and treated with formalin. The treated soil is piled up in a heap and covered with another plastic sheet for 48 hours. Later, the soil is uncovered and stirred frequently to remove the formalin fumes. This soil is fit for casing after about a week when it is free from the smell of formalin.

9.1.12. Maintenance of Environment in Mushroom House

Temperature around 22-24°C till one week after casing is most desirable and subsequent temperature should be 14-18°C and relative humidity of 90 per cent in the mushroom house during entire fruiting period. Initially less fresh air is needed, but during pin-head formation and cropping more fresh air is required.

The genus *Agaricus* has two cultivated species namely *A.bisporus* and *A. biturquis*. The *A.bisporus* is low temperature species and requires 24±1 °C for its vegetative growth and 14 ±18°C for its fruiting. In Haryana it can be grown conveniently during October to Feb. *A. biturquis* grows at higher temperature and is known as temperature tolerant. It requires 4-6 °C higher temperature than *A.bisporus* for its growth and fructification.

9.1.13. Harvesting and Marketing

Mushroom is to be picked when it is button stage (Figure 17). For this, it is to be held between the thumb and first two fingers and rotated gently clockwise and anticlockwise and pulled out with care. Lower portion of the stem, which is covered with casing mixture, is trimmed with a sharp knife followed by washing and drying.

Figure 17: Commercially Grown White Button Mushroom.

9.2. Production of Oyster mushroom

Oyster mushroom which is popularly known as 'Dhingri' in north India is a Basidiomycetous fungus that belongs to the genus *Pleurotus*. It is cellulose loving fungus and grows naturally in the temperate and tropical forests on dead and decaying wooden logs or sometimes on dying trunks of deciduous or coniferous woods in different parts of the world. It may also grow on decaying organic matter. The fruit bodies of this mushroom are distinctly shell or spatula shaped with different shades of white, cream, grey, yellow, pink or light brown depending on the species.

The cultivation of *Pleurotus* species on trees stumps and logs was first described by Falck in the year 1917. Since then much progress has been done in its cultivation. At present it is produced worldwide. The cultivation of *Pleurotus* species, commonly called as Oyster mushroom or Dhingri is gaining popularity because of its diverse ability of growth at a wide range of temperature (15-30°C) and on various agricultural waste materials.

9.2.1. Commonly Cultivated Species

1) *Pleurotus florida*

The colour of this species changes with temperature. At low temperature (about 5°C) the pileus is of light brown colour; at maximum temperatures for the development of fruit bodies (25°C) it turns paler to pallid yellow or white.

2) *P. flabellatus*

It is pure white in colour. Fruit bodies always appear in large clusters. The pileus is having thin flesh with mild aroma. On cooking it gives slightly leathery or fibrous texture, specially the stem portion.

3) *P. sajor-caju*

It is an indigenous species first cultivated by Jandaik in 1974 on banana pseudostems. The fruiting bodies are pale grey to dark gray in colour with good aroma. Temperature requirement is from 25°C to 30°C. At CCSHAU, gamma irradiated strains of *Pleurotus* have been developed which produce fruit bodies at 10-15 °C.

4) *P. cornucopiae*

The colour is dull- ochreous, pallid whitish to light yellow colour and cap 4-10 cm in diameter. It grows well at 15±2°C.

5) *P. ostreatus*

It can fruit at 15°C and is of grey, grey brown or slate grey colour. It spores are known to be causing allergy in many cases.

6) *P. ostreatus* (Grey)

The fruiting bodies are dark slate grey to bluish grey in colour with thick non-fibrous and good aroma of flesh. The gills are white.

7) *P. eryngii*

It is widespread in southern Europe and the areas of Central Asia and North Africa. In India it is found in Jammu and Kashmir. It occurs as parasite on the roots of *Eryngium campestre*. The colour is reddish brown, grey brown to dirty yellow, cap is 4 to 5 cm wide. The weight of the single fruit body is 300-400g.

9.2.2. Cultivation Method and Process

The simple container like polythene bags, polythene tubes, gunny bags, bamboo baskets, wooden trays, plastic sacs and plastic boxes etc. can be used. In all these cases however, care should be taken for the exchange of gases by inserting some perforated tubes in the substrate heap to avoid anaerobic fermentation. Some of the popular methods of cultivation employed are (Figure 18).

1. Bag System

The polythene bags with holes made in them are best containers for oyster mushroom cultivation. Substrate is filled in the polythene bags and spawned simultaneously. The mouth of the bag is tied or folded and kept for spawn running.

Bag System   **Tray System**

Figure 18: Cultivation Methods of Oyster Mushroom.

2. Bed System

The racks made out of bamboo (locally available) material can be utilized for this type of cultivation. The inoculated substrate should be spread on the racks to a thickness of 9 inches and covered with polythene sheet for good spawn run and also to minimize the evaporation of water.

3. Tray System

Wooden plastic trays or empty fruit boxes can be used.

4. Pillar System

A self supporting structure is created by inserting a perforated cement or PVC inside the pillar containing substrate. The perforated tube gives chimney effects for exchange of gases. Polythene tubes with 1-2 ft. diameter, and a height of 5-6 ft. would accommodate a minimum of 25 kg dry straw and yields 20-30 kg of fresh mushrooms.

Cultivation Process

The procedure for oyster mushroom cultivation can be divided into four following steps.

1. Preparation of spawn.
2. Substrate preparation.
3. Spawning of substrate.
4. Crop management.

9.2.2.1. Preparation of Spawn

The spawn preparation technique for oyster mushroom is same as that of white button mushroom. One should have a pure culture for inoculation on sterilized wheat grain. The mycelium of oyster mushroom grows very fast on wheat grain and 25-30 days old spawn starts forming fruit bodies in the bottle itself.

9.2.3. Substrate Preparation

Substrates of Oyster Mushroom

The *Pleurotus* species has the ability to break down cellulose and lignin bearing material without chemical or biological preparation. The straw is soaked overnight in water. This is done to achieve the moisture content of 70 per cent. It also helps in the removal of some surface contaminants on the straw.

9.2.3.1. Sterilization

Sterilization is done by any of the four methods

i) **Chemical sterilization**: It is done with fumigants like formaldehyde and fungicides. 7g carbendazim + 125 ml formalin can be added to 100 ltr. of water for soaking 30 kg wheat straw.

ii) **Steam sterilization**: Steam sterilization is done by injecting steam generated in the boiler to a chamber stacked with substrate material to be disinfected. The temperature is maintained in the chamber at 60-65 °C for 4hr. or at 80°C for 30 minutes.

iii) **Hot water sterilization**: Hot water treatment is given by dipping the substrate in boiling water (80-85°C) for 30 minutes.

iv) **Solar sterilization**: After wetting, the substrate is spread on a clean surface / sheet up to a height of 1 feet and covered with transparent polythene sheet in the sun. It should be in the sunlight for 4 hrs.

v) **Sterile technique**: It is also known as till method. The chopped substrate after soaking in cold water is put in heat resistant polypropylene bags and sterilized in autoclave at 20 p.s.i. pressure for 1-2 hours followed by spawning after cooling under aseptic conditions. This method is more suitable for research work rather than commercial production.

9.2.4. Decomposition or Composting

This method is a modification of composting technique used for white button mushroom. It is most suitable for hard substrate like cotton stalks, maize stalks and leguminous stubbles etc. Both aerobic and anaerobic decomposition of the substrate is suitable for *Pleurotus* cultivation. Composing should be done on a covered area or shed. Chop the substrate into 5-6 cm long pieces. Add ammonium sulphate or urea (0.5-1 per cent) and lime (1 per cent) on dry weight basis of the ingredients. Horse manure or chicken manure (10 per cent dry weight basis) can also be used instead of nitrogenous fertilizers. Addition of lime improves the physical structure of the compost. After mixing all the ingredients sprinkle water till it is completely wet. Prepare a triangular heap of 75-90 cm but not more than 1 meter high. After 2 days of decomposition turning of pile is done adding 1 per cent superphosphate and 0.5 per cent lime. The compost will be ready after 4 days of this turning. It can be spawned as such or used after pasteurization.

9.2.5. Spawn Run

After the operation of spawning is complete, the trays/racks are covered with polythene sheets in order to retain the moisture. In the case of polythene bags the mouth of the bags are tied. Spawned trays/bags are kept in a dark place with minimum air movement. The substrate is kept moist by sprinkling water. The polythene sheets are removed after 15-20 days of spawning or the polythene bags is removed and watering is done once daily. It prefers a little high CO_2 and relative humidity around 20-30°C or depending on the species used for cultivation.

9.2.6. Cropping

Fifteen to twenty days after spawning, there is complete mycelial impregnation in the substrate which becomes white in colour due growth of the fungus. Light is the initiating factor in the development of primordial, therefore needed for at least 15 minutes per day. Temperature and humidity too have an effect on the yield.

9.2.7. Harvesting and Marketing

The mushroom should be harvested when the pileus is about 5-6 cm in diameter and before spores are released. The fruit bodies should be harvested by twisting them, so that broken pieces are not left in the substrate and surrounding fruit bodies are not disturbed. After harvesting the lower portion of the stalk with adhering debris etc. should be cut with the help of clean knife blade. This mushroom may be consumed fresh or dried in sun or a mechanical dehydrate at 50-60°C. This mushroom can be powdered and stored in air tight containers and used whenever required.

9.2.8. Precaution While Growing Oyster Mushroom

Oyster mushrooms produce millions of spores which can be easily seen as spore clouds in the cropping rooms in early morning. Many times growers working in cropping rooms complain of headache, fever, joint pains, nausea and coughing due to *Pleurotus* spores. Mushroom pickers are advised to open the doors and ventilators or switch on exhaust fans 2-3 hours before entering the cropping room. Pickers should use respiratory masks in growing rooms and change clothes after coming out from growing rooms.

9.3. Production to Milky Mushroom (*Calocybe indica*)

Mushroom production represents one of the most commercially important microbial technologies for large-scale recycling of agro-wastes, available in plenty in an agriculture country like India. It relieves the pressure on arable land because its cultivation done is indoors. Mushroom supplement compliment the nutritionally deficient cereals, and are regarded as highest producers of protein per unit area and time. With the emphasis being given to the horticultural cash crops, mushroom cultivation is gaining importance and becoming popular in our country as production increased up to 70,000 tonnes per annum. There virtually had been a revolution in mushroom cultivation but in order to bring our country on the mushroom map of the world serious affords are needed to perfect the production technologies of newer edible mushrooms including *Calocybe indica*.

The milky mushroom or white milky mushroom (*Calocybe indica*) belongs to the family Trichalomatceae of the order Agaricales. It was first reported from India in 1974 and was first cultivated in 1976 by Purkayastha and Chandra. This mushroom is known by diverse names in different parts of India and most common name is Dudh Chhata / Safed Dudhiya Mushroom. Its robust size, milky white colour, delicacy and long shelf life has attracted attention of both consumers and prospective growers. Subsequently, its nutritional requirement and nutritional value were also determined which are at par with other edible mushroom. Nutritional value of *Calocybe indica* dried sporocarp is having protein 17.69 per cent, fat 4.1 per cent, crude fibre 3.4 per cent and carbohydrate 6.42 per cent. Natural sporocarp contains 4 per cent sugar, 2.9 per cent starch and 7.43 per cent ash. In addition to this, it has most of the mineral salts required by the human body, such as potassium, sodium, phosphorus, iron and calcium. Due to its alkaline ash and high fibre content, it is highly suitable for people with hyperacidity and constipation. As consumers become more aware of the additional culinary characteristics offered by this mushrooms, the demand of *Calocybe indica* will increase. Its production cost is much lower than that of *A.bisporus* in view of its simple cultivation technology. It is cultivated on large scale in southern states like Tamilnadu, where its production during last three years is increasing. As it can be cultivated in the temperature range of 25-35°C and has good biological efficiency (60-70 per cent) in Northern India it can easily fit in the relay cropping of mushrooms when no other mushroom can be grown at higher temperature with more than 60 per cent biological efficiency.

Straw is chopped in small pieces (2-4 cm size) and soaked in freshwater for 8-16 hours. This period can be reduced when pasteurization is to be done by steam. Main purpose of soaking is to saturate the substrate with water. It is easier to soak if straw is filled in gunny bag and dipped in water.

9.3.1. Pasteurization

The purpose of pasteurization is to kill harmful microbes. This can be achieved in three ways.

(i) Hot Water Treatment

Water is boiled in wide mouth container and chopped wet straw filled in gunny bag is submersed in hot water for 40 minutes at 80-90°C to achieve pasteurization. This is very popular method particularly with small growers.

(ii) Steam Pasteurization

Wet straw is filled inside insulated room either in perforated shelves or in wooden trays. Steam is released under pressure from a boiler and temperature inside substrate is raised to 65°C and maintained for 5-6 hours. The chamber used for casing soil pasteurization for button mushroom cultivation will be ideal. Air inside the room should be circulated to have uniform temperature in the substrate.

(iii) Chemical Pasteurization

Chemical pasteurization of the substrate is done using formalin and bavistin as in the case of oyster mushroom. Soak 10-12 kg of fresh paddy straw/wheat straw in 90 litres of water in a wide mouthed container. Dissolve 7.5 g bavistin (carbendazim) and 125 ml formaldehyde (commercial) in 10 lts of water in a bucket by stirring for a few minutes and pour it on the substrate to make it 100 litres of water in container and cover the mouth of the container with polythene sheet and leave it overnight (minimum 16 hrs) for effective pasteurization.

9.3.2. Sterilization

Substrate is filled in polypropylene bags (35-45 cm, holding 2-3 kg wet substrate) and sterilized at 15 lbs p.s.i. for 1 hour. Once pasteurization/sterilization is over straw is shifted to spawning room for cooling, bag filling and spawning.

9.3.3. Spawn Preparation

The cereal grains like wheat, sorghum, pearl millet etc. can be used as substrate for spawn preparation. In Haryana healthy wheat grain are boiled for 20 min, taking care that they do not split. Then after straining and drying, grains are mixed with 2 per cent gypsum and 6 per cent calcium carbonate (wet wt. of the grains). About 250 g of grains are taken in each half litre milk bottles/empty glass glucose bottles or polypropylene bags and sterilized for 2 hrs at 20 lb. pressure psi. After cooling, each bottle/polypropylene bag is inoculated with mycelial bit (7-10 days old culture) of *C.indica* and incubated for 15 days at 30°C and provided light (6-7 hr. /days). The contents of bottles/polypropylene bags are shaken time to time for uniform growth of mycelium and to avoid clumping of the grains.

9.3.4. Substrate Preparation

Generally, wheat or paddy straw is used for cultivating this mushroom. Different other substrates have also been tried for its cultivation; such as sorghum stalks, brassica straw, sugarcane bagasse, black gram, hay etc.

9.3.5. Spawn and Spawning

About 15 to 20 days old spawn is used for spawning. Spawning can be done by broadcasting and layering methods. Nevertheless, layering method is better than the others. Three layers are made with 4 per cent of grain spawn in bags or beds. Optimum temperature for mycelial growth ranges between 25-35°C. The spawn run is completed in 15-20 days depending on the environment prevalent in the mushroom house.

9.3.6. Casing

After complete spawn run casing is needed for fruit body initiation. When the substrate is completely covered with mushroom mycelia casing should be done. Various casing materials and their combination can be utilized such as garden soil and FYM (1:1), cow dung patties (2 years old), biogas slurry, FYM with various combinations. At CCS Haryana Agricultural University, Hisar, burnt rice husk alone or in combination with FYM /field soil gave very encouraging results as casing material. The thickness of casing should be kept at 1-5″.

9.3.7. Fruiting

Mycelial strands are usually observed on the surface of casing layer within 3-5 days. After 10-12 days of casing, fruit primordia are formed and within 5-6 days mature fruit bodies are seen which are ready to harvest (Figure 19)

Figure 19: Commercially grown Milky Mushroom.

9.3.8. Management

It requires high temperature, high relative humidity, diffused sunlight and good aeration. Fruiting is affected in absence of these conditions. For fruiting, temperature ranges of 25-35°C are required. The pH of casing material should be around 7.0 and relative humidity should be more than 70 per cent. Diffused sunlight is needed for a good crop.

9.3.9. Harvesting

First flush occurs 3-4 weeks following spawning; generally mature mushrooms (with well differentiated pileus, gills and stipe) are picked up gently without disturbing the smaller ones by twisting so that the stubs are not left on the straw. Fresh mushrooms after cutting the lower portion of stalk are packed in perforated polythene bags for marketing. The mushrooms can be cut in the form of chips and dried in sun or hot air oven at 40-45°C without any changes in colour. This dried produce can be stored for one year after sealing in polythene bags. It can be rehydrated in water within 10-15 minutes, giving 80-90 per cent of original weight. About 700g of fresh mushroom can be harvested per kg of the substrate.

9.4. Cultivation of Paddy Straw Mushroom (*Volvariella* sp.)

Paddy straw mushroom (*Volvariella volvacea*) is the most popular mushroom in the rural areas of the Philippines and was commonly known to Filipinos even before the generation of a technology for its artificial cultivation was launched in the country. This mushroom is called kabuteng dayami or kabuteng saging by Filipinos since it naturally grows on paddy straw and decomposing piles of banana leaves and pseudostems. Unlike the commercial production of paddy straw mushroom in neighboring countries such as China, Vietnam, Indonesia, and Thailand, production in the Philippines is still a backyard undertaking. Attempts to upgrade its production were made in the early part of the 1990's with financial assistance from international donors (Tharun, 1993; Ferchak and Croucher, 1993). Still, paddy straw mushroom production has been out paced by the newly introduced *Pleurotus* mushrooms, despite the abundance of substrates, availability of technology and high market demand.

The cultivation of this mushroom was first tried at Coimbatore and since then its cultivation has been taken over at other places also. It grows at high temperature between 30 and 45° C. It can therefore, also be cultivated during the summer months. A variety of materials have been tried for cultivation, for example, cotton waste, paddy straw etc. In India, three species of *Volvariella* are grown, namely *V. diplasia*, *V. volvacea* and *V. esculenta*. *Volvariella volvacea* is the most commonly grown mushroom out of these species.

Though indoor technology had already been introduced, most growers still adopt the traditional method due to its simplicity and low input. Maintenance of environmentally controlled conditions is one of the necessary factors to attain stable and relatively higher production, and would ultimately ensure the regular availability of paddy straw mushroom in the market.

9.4.1. Paddy Straw Mushroom Production Technology

For fruiting body production technologies that are being adopted by the growers include: Outdoor and indoor techniques.

9.4.1.1. Outdoor Cultivation of Paddy Straw Mushrooms

The traditional outdoor method uses the bed-type approach and utilizes a number of agricultural wastes like dried paddy straw, rice stubbles, water lily, banana leaves, and stalks. The use of these wastes as mushroom substrates depends on their local availability. The following is a description of the step by step procedure for the preparation of mushroom cultivation beds

9.4.1.1.1. Site Selection and Preparation

The selected site should preferably be under trees with a wide canopy. Growers should choose an area that is free from potential insect pests such as ants, termites and rodents. In order to ensure that the selected site is pest free, growers can spread rice hulls onto the area and burn them until they turn into ashes. This physical method of eliminating pests also reduces the occurrence of soil-borne pathogens.

9.4.1.1.2. Collection and Preparation of Bedding Materials

The bedding materials collected from the field should be sun dried. Growers should trim and bundle the substrates into bundles twelve inches long with a diameter of two inches. The bundled substrates should be soaked for twelve hours and washed with clean water.

9.4.1.1.3. Layering of Bundled Substrates into Bed and Spawning

The bundled substrate should be drained of excess water to attain 65 per cent moisture content. Growers should pile the bundled substrates one after the other into the bed forms. On top of every layer, spawn should be sprinkled thinly over the bundled substrates. An ideal bed size consists of six layers and has a length of three meters.

9.4.1.1.4. Incubation and Fruiting

A plastic sheet should be used to cover the entire mushroom bed. This sheet maintains the appropriate temperature for the mycelial ramification (30-35°C) and fruiting body formation (28-30°C). It usually takes 10-14 days before the first flush of marketable fruiting bodies (button stage) come out from the edge of the mushroom bed.

9.4.1.1.5. Harvesting

With bare hands, growers should harvest the button stages of *V. volvacea* by simply pulling the cluster out from the bed.

9.4.1.2. Indoor Cultivation of Paddy Straw Mushrooms

A more improved technology that is now gaining interest among potential growers is the indoor production technology. This method, which utilizes paddy straw as the main substrate, has three salient features: composting, pasteurization, and cultivation inside a mushroom house.

The cultivation of mushrooms inside a growing house allows for the control of the fluctuations of temperature and relative humidity which may be hazardous to the mycelial growth and fruiting body production. In the Philippines, two methods of indoor cultivation have been developed and introduced. The key features of both are similar, but the manner by which spawn is inoculated into the substrates differs. The indoor cultivation is a standard method that is also used abroad, and features the actual spawning on the mushroom beds. The other method promotes the use of wooden shelves or crates which facilitate the easy handling of substrates.

The following section describes the step by step procedures for the indoor cultivation of *V.volvacea*.

9.4.1.2.1. Soaking

Rice straw of any type can be used as substrate for the indoor cultivation of *V.volvacea*. Rice stubbles could also be used. The rice straw should be soaked for 12 hours in clean water. This procedure loosens the substrates as a prelude to composting.

9.4.1.2.2. Composting

The previously soaked substrates should be piled up and sprinkled with 1 per cent molasses and 0.5 per cent complete fertilizer. Growers should cover the pile of substrates with plastic sheets and compost the pile for 14 days. On the seventh day, the partially composted substrates should be turned with a spading fork in order to ensure even composting. At this stage, the population of thermophilic decomposers starts to pile up. Growers should now add 1 per cent agricultural lime, replace the plastic sheets and continue the composting process until completing the required fourteen day composting period.

9.4.1.2.3. Crating and Steaming

Growers should dispense the composted substrates on 12 x 24 x 18 inches wooden crates that are open on all sides. Moisture content of the substrate should be 65 per cent (no drippings of water when squeezed between fingers). Growers should make sure that the substrates are compactly placed inside the wooden crates, and should deliver the crated substrates into the steaming room by piling them one on top of the other. Growers should then start introducing the steam into the mushroom house. Steaming usually lasts from four to six hours, and the temperature should maintained at 60-80°C.

9.4.1.2.4. Production of Spawn

A number of locally available substrates are being used as spawning material for paddy straw mushroom. In Northern Luzon for instance, tobacco midrib, a waste product of the cigar industry, is being used by the spawn producer of Pangasinan in Northern Luzon. Tobacco midrib that has been soaked in water for three days and later washed and air dried is mixed with sawdust. The mixed formulation is then placed in empty mayonnaise bottles and sterilized by autoclaving. In Central Luzon, the Center for Tropical Mushroom Research and Development at the Central Luzon State University developed and introduced the use of rice hull, a waste material from rice milling. Rice hull is moistened and mixed with 10 per cent of either corn meal or rice bran and dispensed in heat resistant polypropylene bags and microwaveable plastic trays. In other areas of the country where leaves of leguminous trees like Gliricidia and Leucaena are abundant, dried leaves of these trees are soaked for three days and later air dried. The leaves are then mixed with sawdust and rice bran at a rate of seven parts leaves, three parts sawdust and one part rice bran. Coffee hulls are also being used in areas where coffee is grown. Moisture content of all the preparations is 65 per cent.

9.4.1.2.5. Spawning the Substrate

The next morning after steaming the substrates, growers should check the temperature of the steamed substrates. The temperature should be 30°C in order not to harm the mycelia of *V. volvacea*.

9.4.1.2.6. Incubation and Fruiting

In order to encourage mycelial proliferation of *V. volvacea*, the mushroom house should be sealed. During this stage, it is very important to maintain the desirable temperature for mycelial ramification (30-35°C) with no ventilation and light. The spread of mycelia takes from seven to ten days after spawning. After this period, growers should check the status of the substrates. Fruiting initials should start to appear. At this point, the temperature should be lowered from 35 to 28°C. This can be done by sprinkling clean water on the floor of the mushroom growing house. Three to five days after the appearance of these fruiting initials, the first harvest of the button stages of *V. volvacea* can be performed (Figure 20).

9.4.2. Nutraceutical Benefits

Though paddy straw mushroom is known primarily for table consumption due to its nutritional content, its use as a functional food has started to be recognized. A number of studies on its immunobiological activities have been reported (Kishida *et al.*, 1992; Kishida *et al.*, 1989; Misaki *et al.*, 1986 and Sone *et al.*, 1994). Thus, its additional use as a nutriceutical could be an additional factor in marketing this type of mushroom.

Figure 20: Commercially Grown Paddy Straw Mushroom.

9.4.3. Zero Farm Wastes Technology for Paddy Straw Mushroom Spent

Paddy straw mushroom cultivation utilizes large volumes of paddy straw as substrates for fruiting body production. Hence, tons of mushroom spent are also generated which results in the accumulation of wastes in the form of mushroom spent. If improperly disposed, these wastes might pose environmental hazards. Traditionally, the mushroom spent of paddy straw mushroom is burned in order to get rid of contaminants. The spent from paddy straw mushroom production can further be efficiently utilized to harness its full potential for food production. It has shown promising results as potential substrates for *Pleurotus, Auricularia, Ganoderma* and *Collybia*, fishpond fertilizer for tilapia (*Oreochromis niloticus*) and feed for broiler chickens (Reyes and Abella, 1997; Abella *et al.*, 1996; Divina *et al.*, 1996a and b; Reyes and Abella, 1993).

9.5. Cultivation of *Llentinus edodes*: Shiitake Mushroom

Lentinus edodes, the shiitake mushroom is now common in Asian countries. *L. edodes* was most often sold in the United States, but because cultivation of this species is now common fresh shiitake is now available at the supermarket. The cultivation of *L. edodes* first began in China about 1100 AD (Nakamura 1983, Royse *et al.*, 1985, Chang and Miles 1987, 1989). It is believed that shiitake cultivation techniques developed in China was introduced to the Japanese by Chinese growers. The shiitake has the distinctive advantage of a much longer shelf life because they are more commonly sold dried. It is liked by consumers because of its unique taste and flavour and presence of a chemical which reduces plasma cholesterol level.

9.5.1. Cultivation Technology

Lentinus edodes grows in nature on the dead wood of a number of hard wood trees mainly *Quercus* spp. (Oak),*Castenopsis* spp., *Lihocaripus* spp., *Carpinus* spp., *Elacocarpus* spp. and *Betula* spp. This mushroom is cultivated throughout the year by adopting improved cultivation techniques either those which can fruit at or above

20°C or between 10-15 °C and those fruiting around 10 °C. The growing technology consists of following steps:

1. The log preparation
2. The spawn preparation
3. The spawning of logs
4. The crop management

The Log Preparation

Lentinus edodes mainly grows on dried wooden logs absorbing nutrient from the cambium. Although it grows on any size and age of logs. The log 9-18 cm diameter and from 15-20 years old trees are most suitable for cultivation. December and January to early March when logs contain highest amount of carbohydrates and other organic substrate are the most suitable period for its cultivation. The log should contain a moisture content of 45-55 per cent with pH between 4.5 to 5.5. To bring optimum moisture level that is 40-45 per cent the filled logs are left as such for 25-45 days.

Spawn Preparation

Spawn has categorized into two types namely: 1. Sawdust spawn 2. Wood plug spawn

1. Saw Dust Spawn

It is prepared from using any of the following formulae.

Formulae-1

Saw dust	–	65 per cent
Wheat bran	–	15 per cent
Used tea leaves	–	20 per cent
Water	–	65 per cent

Formulae-2

Saw dust	–	78 per cent
Sucrose	–	1.0 per cent
Wheat bran	–	20 per cent
Calcium carbonate	–	1.0 per cent
Water	–	65 per cent

Formulae-3

Saw dust	–	800gms
Rice bran	–	200gms
Sucrose	–	30 gms
Potassium nitrate	–	4gms
Calcium carbonate	–	6gms
Water	–	2 liters

After properly sieving, the saw dust is thoroughly mixed with water by removing bigger size of wood particles and other impurities. Ooze out one or two drops of water by squeezing and filled in wide mouthed bottles or in polypropylene bags. Use rod to make one inoculation hole into centre of the substrate. Autoclaved at 20lbs pressure for 2 hrs after plugging with non absorbent cotton and covering with aluminum foil. 10 days old culture of actively growing mycelium is inoculated aseptically and incubated for 30 days at 24±2°C.

2. Wood Plug Spawn

Wood plug spawn is prepared by incubating mycelium on small wedge shaped or small cylindrical wood species. They are ready for the inoculation when the fungal mycelium impregnates the wood species.

Spawning of Logs

The shiitake mycelium commonly grows between the temperature ranges from 5-30°C but the most optimum temperature has been observed to be ranging between 20-26°C. To avoid competitors mould, the spawning should be done at low temperature ranges from 14-20°C. For spawn inoculation, small holes measuring 1.0×1.0 cm in size and 1.5-2 cm deep should be made at a distance of 20-30 cm on long axis. Saw dust spawn is filled in the holes or wood plug spawn is inserted by cutting out similar size pieces. The saw dust spawn should not be kept tightly pressed. It should be kept soft and holes sealed with paraffin wax. The spawning should be mostly done in aseptic condition.

Crop Management

Inoculated logs are kept in open at a place where the physical conditions are most favourable for the mycelium growth. With minimum exposure to light the inoculated logs are kept in a flat pile and covered with either straw or gunny bags to prevent excessive water loss of the logs. Depending upon the culture strain and type of wood used the vegetative growth in the logs. Induction of fruit body requires temperature shock or temperature drop, high humidity and enough light. To induce fruiting following precautions are to be taken.

1. For temperature shock, the logs are either sprayed with cold water or immersed in a tank of cold water.

2. During summer when the logs are immersed in cold water they should be kept for 24 hrs at 15-18°C, while during winter they should be kept for 2-3 days at 10-15°C.

3. The cropping area is kept moist to maintain high relative humidity and the logs are then leaned against the supports.

4. The temperature should be maintained between 15-20°C and then humidity around 80-90 per cent.

5. Fruit bodies are harvested by first pressing and then twisting it gently.

6. Mushrooms are harvested upto three times and after a rest for 30-40 days they are again watered to get more mushrooms. It can be repeated for 3-4 times per year and these logs will produce crop upto 4-6 years.

9.5.2. Synthetic Cultivation

The cultivation of *L. edodes* by log method requires 3-5 years. It cannot be produced easily and being very costly. The commercial cultivation is carried out by using saw dust of oak (*Quercus* sp.), maple (*Aser* sp.) or berch (*Betula* sp.).

For substrate preparation various formulation are in use which are given as under.

Formula-1

Saw dust	–	80 per cent
Rice bran	–	20 per cent
Water content	–	65 per cent

Formula-2

Saw dust (maple and birch)	–	80 per cent
Millet	–	65 per cent
Wheat bran	–	80 per cent
Calcium	–	20 per cent
Saw dust	–	65 per cent

Formula-3

Hard wood saw dust	–	89.2 per cent
Rice bran	–	10 per cent
Calcium carbonate	–	0.2 per cent
Water content	–	60 per cent

Formula-4

Corn cobs	–	40 kg
Saw dust	–	10 kg
Wheat bran	–	12.5 kg
Cane sugar	–	1kg
Pectin	–	15g
Urea	–	20 g

Formula-5

Paddy straw	–	50 per cent
Wheat straw	–	20 per cent
Saw dust	–	20 per cent
Cane sugar	–	1.3 per cent
Calcium carbonate	–	1.5 per cent
Citric acid	–	0.2 per cent
Calcium sulphate	–	0.5 per cent

Figure 21: Commercially Grown Shiitake Mushroom.

The water content of the substrate should be adjusted to 60-65 per cent and pH between 5.5-6.0. Soluble ingredients such as citric acid and sugar etc. are usually first dissolved in water before mixing. Saw dust should be soaked at least for two days and rice straw for three hours. All the ingredients are thoroughly mixed.

9.5.2.1. Bag Filling Technique

The bags should be filled at the rate of 1.5 to 4 kg per bag immediately after mixing and wetting the substrate, otherwise fermentation and contamination may cause deterioration and damage to the substrate. The polypropylene bags are first loosely filled and later the substrate is pressed by putting gentle pressure. It gives the cylindrical shape to the bags. The polypropylene bags are then autoclaved for 1 hr at 121°C. After cooling down, the bags are inoculated with spawn 2 to 5 per cent and kept at 20 to 25 °C in dark for 45 to 60 days to complete the mycelial growth.

The polypropylene bags are removed when the mycelium in the bag is matured. By lowering the temperature ranges from 16 to 18 °C or by soaking in cold water the compressed blocks are treated for fruiting. The primordia appear within 3 to 4 weeks time and mushroom develops in another 7 to 10 days. A good quality mushroom crop can be harvested when the temperature in the cropping rooms ranges between 15 to 25 °C. The relative humidity should be maintained about 85 to 90 per cent. There is no need of light during primordial formation but it plays a role in the subsequent fruit body formation. Some of the strains of *L. edodes* with low light intensity (120-200 lux) gave higher yield while with medium light intensity (500-600 lux) yielded poorly. On the other hand low light intensity lengthens their cultivation period while medium light intensity stimulates them to fruit more rapidly.

9.5.2.2. Harvesting, Drying and Storage

It should be done at an early stage. Normal yields are 15-30 per cent of the wet weight of substrate. Shiitake mushroom can be eaten fresh, dried, canned or pickled in vinegar. The dried shiitake should have 10- 13 per cent moisture.

9.6. Cultivation of Black Ear Mushroom (*Auricularia* species)

Genus *Auricularia* belongs to a group of fungi characterized by gelatinous bodies and different species of *Auricularia* are known as black ear mushroom. Among the cultivated mushroom, black ear mushroom has the oldest record of cultivation by Chinese dating back to 600AD. In some of the Asian countries, at present its cultivation has become a major business. *Auricularia* spp. production now represents about 7.9 per cent of the total cultivated mushroom supply world-wide and ranks fourth among the different cultivated species of mushroom. All the species of *Auricularia* so far are edible. Among the cultivated species *A. auricularia* and *A. fuscosuccinia* are light coloured and small, while *A. polytricha* are dark coloured, large and hairy and do not turn slimy on cooking. Black ear mushrooms are believed to have medicinal importance and are used to cure sore throat, anaemia and certain digestive disorders especially piles and chronic constipation.

9.6.1. Cultivation Technology

The substrate may be composted for up to 5 days or used directly after mixing. In either case, the mixed substrate (about 2.5 kg wet wt) is filled into heat resistant polypropylene bags and sterilized (substrate temperature 121°C) for 60 min. Composted substrate is prepared by mixing and watering ingredients [sawdust (78 per cent):bran (20 per cent):calcium carbonate (1 per cent):sucrose (1 per cent)] in a large pile. The pile then is covered with plastic and turned (remixed) twice at two-day intervals. For direct use of substrate, a mixture of cottonseed hulls (93 per cent), wheat bran (5 per cent), sucrose (1 per cent), and calcium carbonate (1 per cent) is moistened to about 60 per cent moisture and then filled into polypropylene bags.

After the substrate has cooled, it is inoculated with either grain or sawdust spawn. The spawn then is mixed into the substrate either mechanically or by hand. After the mycelium is allowed to colonize the substrate (spawn run), temperatures for spawn run are maintained at about 25°C ± 2°C for about 28 to 30 days. Light intensity of more than 500 lux during the spawn run may result in premature formation of primordia. Temperature, light intensity and relative humidity all interact to influence the nature and quality of the basidomata.

9.6.1.1. Cultivation on Logs

1. The logs with moisture content of 50-65 per cent and with enough sugar accumulation are considered to be most suitable for cultivation.

2. The trees should be normally 6-10 years old and a log length of one meter with diameter 3-6 cm is preferred.

3. To inoculate spawn, the holes are bored around the logs by means of a chisel or with an electric drill and filled with saw dust spawn. The holes are then plugged with bark and sealed with molten wax.

4. The logs are inoculated by laying them in shade with optimum temperature of 24-26°C.

5. To maintain proper moisture content, the inoculated logs are laid crosswise or stand upright in the yard and covered with plastic sheet.

6. The logs are ready for cropping after about 60 days and transferred to the cropping yard where temperature 22-25°C and moisture content 80-90 per cent are maintained by frequently spraying logs with water.

7. Mushrooms can be harvested for about 30 days and if the logs are protected properly, harvest can be taken from these logs every year in the natural growing seasons. The total yield of cropping is reported to be 12-15 per cent of the original weight of the log.

9.6.1.2. Cultivation in Bags

Due to scarcity of wood as well as ease in handling *Auricularia* cultivation is done in polypropylene bags. For cultivation the substrate like wheat or paddy straw is used. The straw is soaked in water overnight for 16-18hrs, after cutting them into small pieces of 4-6 cm. After draining excess water, the soaked straw is supplemented with 4-5 per cent of wheat or rice bran. The mixture is filled in the polypropylene bags @ 4 kg per bags and autoclaved at 15lbs pressure for 1-2 hrs. Spawn is added to the substrate aseptically after cooling @ 2 per cent and incubated at 25-28°C for 25-28 days. Slitting of bags for fruiting prevented drying out of substrate which was found to be more beneficial than completely removal of bags.

For fruit body development and maturation temperature, humidity, light, aeration and watering are pre-requisite. Give daily spray, 1 hour diffused light and aeration.

9.6.1.3. Management

Maintenance of proper hygienic conditions are necessary because of addition of supplement like wheat or rice bran which may attract moulds. Development of abnormal fruit bodies with long stalks and small pileus is due to the presence of carbon dioxide. The absence of light also produces abnormal under developed mass of fruit bodies.

Figure 22: Commercially Grown Black Ear Mushroom.

9.6.1.4. Harvesting

Auricularia polytricha mushroom harvested from wood logs has been observed to be tough in texture, less in colour attraction with larger production period in comparison to *Auricularia polytricha* mushroom harvested from substrate bags. The fruit bodies of this mushroom can easily be sun dried resulting in dry matter of 8-12 per cent. This mushroom has a special quality of retaining its characteristics crispness on cooking.

9.7. Cultivation of Winter Mushroom (*Flammulina velutipes*)

Worldwide production of winter mushroom has increased from about 143,000 metric tons in 1990 to about 285,000t in 1997. Japan is the main producer of *F. velutipes*. *F. velutipes* forms small fruit bodies and is liked due to its delicious taste. This mushroom usually appears in winter season in Japan. Therefore, it is called winter mushroom. Sawdust cultivation is preferred because quality mushrooms are produced through this method.

9.7.1. Cultivation Technology

9.7.1.1. Substrate Preparation

The cultivation process from spawning to cropping usually takes about 3 months. During substrate preparation the substrate is composed of sawdust preferably taken from broad leaf trees and rice bran in the ratio of 4:1. The ingredients are thoroughly mixed, water is added to produce a moisture content of 58-60 per cent and the medium is mixed again. It is now filled in polypropylene bottle of 800-1000ml capacity. About 540g of substrate is filled in a bottle. Polypropylene bottles or bags are filled with substrate mixture and autoclaved for an hour at 121°C.

9.7.1.2. Spawn and Spawning

After the substrate has cooled to 20°C, it is inoculated with sawdust spawn. The inoculated substrate is placed in an incubation room at 20-25°C for mycelial growth. When the mycelium covers 90 per cent of the substrate surface after 20-25 days, the upper layer which consists of sawdust spawn is then removed. These bags are now placed in the dark at 8-12°C and 80-85 per cent humidity to stimulate primordial formation.

9.7.1.3. Fruit Body Formation

The fruit bodies will appear after 12-15 days. The temperature is lowered to 5-8 °C. The fruiting bodies formed are stiff, white and relative dry at this temperature and with some ventilation. When the stipes has reached to a length of 2cm, a thick waxed paper is wrapped around the neck of the bags to hold mushroom upright. When the fruit bodies are about 2-3 cm long, a support in the form of string or polythene stripes should be tied around the stipes so as not to allow the fruit bodies to drop down on elongation.

9.7.1.4. Harvesting and Marketing

When the stipes are about 8-10 cm long the fruit bodies are harvested. It should be carried out by twisting the fruit bodies in an anticlockwise direction. After

Figure 23: Commercially Grown Winter Mushroom.

harvesting, the lower portion with adhering debris should be cut with the help of a clean knife. About 100-150g mushroom per bottle/bag may be obtained in the first flush followed by considerable reduction in second phase.

10

Mushroom Diseases, Competitor Moulds and their Management

Like other crops, mushrooms are also affected by a large number of biotic and abiotic factors. Among the biotic agents, fungi, bacteria, virus, nematodes and insect pests cause considerable damage to mushroom directly or indirectly.

Hygiene is most important in mushroom cultivation. Unhygienic conditions in and around mushroom growing houses may cause infection of diseases and competitor moulds, which may result in yield losses and sometimes complete crops failure. Following are common diseases/competitor moulds of mushroom.

10.1.Fungal Diseases

10.1.1. Dry Bubbles (*Verticillium fungicola*)

Dry bubble is the most common and serious disease of mushroom. If it is not controlled, disease can totally destroy a crop within 2-3 weeks.

Symptoms

Whitish mycelial growth is initially noticed on the casing soil which turns grayish yellow. At early stage infection, onion shaped mushrooms are produced. Sometimes they appear as small undifferentiated masses of tissue. When affected at later stage, light brown spot on the caps and downwards splitting of stem resulting in shattered appearance. Affected mushrooms are greyish in colour. On fully developed sporophores, it produces localized light brown depressed spots. Adjacent spots coalesce and form irregular brown blotches. Diseased caps shrink in blotched area, turn leathery, dry and show cracks. Infected fruit bodies are malformed, onion shaped and become irregular.

Verticillium is carried to mushroom farms through casing soil and its spread is by equipments, hands and clothings. Phorid, scarid flies and mites are also known to transmit this disease from infected to healthy mushroom. *Verticillium* is soil borne and spores survive in the moist soil for 12 months. The optimum temperature for disease development is 20°C. The period from infection to symptom expression is 10 days and 3-4 days for cap spotting.

Management

Use of sterilized casing soil and proper disposal of spent compost are essentials to avoid primary infection. Maintain proper temperature, ventilation and humidity. The diseased mushrooms should be picked up carefully and destroyed. Sanitary conditions should be maintained in the mushroom houses and when disease appears then temperature should be lowered. Spray Bavistin @ 0.05 per cent at 10 days interval.

10.1.2. Wet Bubble (*Myccogone perniciosa*)

It is one of the serious diseases in almost all the major mushroom growing countries of the world. The disease was first described from Paris in 1888 and now occurs in most of the countries. In India, this disease was reported for the first time in 1978 from Himachal Pradesh and now is prevalent in Haryana, Punjab, J and K, UP and Maharashtra.

Symptoms

Fungus covers the mushroom with a dense white mat of mycelium leading to reduction in yield. In early stages, the mushroom has a swollen stalk and a small cap.

The casing soil, unpasteurized compost, implement and workers are the sources of infection. The disease spread can be air-borne, water borne or by mites and flies. Temperature above 17°C is favourable for disease development.

Management

Diseased mushrooms should be removed and destroyed. The infected area should be sprayed with 0.2 per cent Dithane Z-78 or Benlate 0.05 per cent or carbendazin 0.05 per cent.

10.1.3. Cob Web (*Dactylium dendroides*)

Cobweb appears first as small white patches on the casing soil and later on spread to the mushrooms. White mycelium covers the stipe, pileus and gills, eventually resulting in decomposition of entire fruit body. In severe attacks, a dense white mould develops over casing. The white colour turns pink or even red with age. The infected mushrooms become discoloured and soft. At later stage such mushrooms are engulfed in a cottony ball of mycelium. On removal of mycelium from affected mushroom, drops of dark brown coloured fluids exudes emitting bitter foul smell. The high temperature and relative humidity encourage the disease.

Management

Proper sterilization of casing soil is significantly important. Infected portions and its surroundings should be removed. Hygienic conditions should be maintained to prevent the spread of the disease. Spray Dithane M-45 @ 0.02 per cent and reduce the humidity to 80 per cent.

10.1.4. Brown Plaster Mould (*Papulospora byssina*)

This competitor mould has also been reported from India causing 90-92 per cent yield losses and is prevalent throughout the country.

Symptoms

White patches on the exposed surface of compost and casing soil in racks/bags due to moisture condensation on sides of bags/racks are seen. Compost usually turns to rust colour. Too wet compost, high temperature (28-30°C) during spawn run and cropping at more than 18°C encourage infection.

Management

Composting should be made carefully, having sufficient gypsum and moisture contents. Treat the infected patches with 2-4 per cent formalin for its control. Fungicides like carbendazim, vitavax, Dithane M-78 and Captan are effective for the control.

10.1.5. Green Mould (*Trichoderma viride*)

Symptoms

Green mould symptoms appears in compost, on casing soil, spawn bottles and grains after spawning. Pure white growth of mycelium appears on the surface of the compost which looks like mushroom mycelium. Later this turns green due to heavy sporulation of fungus. When it appears on the casing soil it checks the pin formation.

This fungus grows on decomposed organic matter and dead mushroom tissues. Improper pasteurization of compost and high humidity are also responsible for the spread of this disease. Spores are spread by air, water and handling.

Management

Proper hygiene conditions should be maintained to prevent the spread of the disease. Sterilize the implements before use. Remove dead buttons immediately. Spray Bavistin @ 0.05 per cent at 7 days interval.

10.1.6. False Truffle (*Dichliomyces microsporus*)

This is the most dreaded competitor in mushroom beds. It causes serious losses to mushroom crop when the compost temperature is above 27°C. Incidence of false truffle in *A. bisporus* grown under natural climatic conditions has been reported from Himachal Pradesh, Punjab, Haryana and U.P.

Symptoms

Small wefts of white cream coloured mycelium in compost and casing soil and more in the layer between compost and casing mixture. The mycelial growth become thicker and develops in to whitish solid, wrinkled rounded to irregular fungal masses. At later stages they become pink, dry and reddish and finally into powdery mass emitting chlorane like odour. This fungus does not allow mushroom mycelium to grow and compost turns dull brown. The yield of the mushroom crop is reduced significantly.

The source of infection is casing soil and pathogen survives in wooden trays from previous crops. Ascospores can survive for 5 years in soil and spent compost and mycelium for 6 months. This is the major source of infection.

Management

Compost should be prepared on *pucca* flour. During composting there is a rise in temperature which activates the ascospores present in soil. Pasteurization of compost under optimum condition completely removes the truffle inoculums. Truffles infected buttons should be picked and buried before the fruit bodies turn brown and spores are ripe. Material used for making racks/mushroom houses like bamboo etc. should be checked by treating the patches with formaldehyde (2 per cent) solution.

10.1.7. Olive Green Mould (*Chaetomium olivaceum, C. globosum*)

Its occurrance was reported from HP initially and later on observed from other states also.

Symptoms

Grayish white mycelium in the compost or growth on the compost surface after spawning, which changes later in to olive green. Due to this spawn growth is delayed. Infection starts due to improper pasteurization and high temperature in absence of adequate fresh air.

Management

Proper pasteurization should be done. Spraying of Dithane Z-78 @0.2 per cent should be done in the affected portions.

10.1.8. Ink Caps (*Coprinus* spp.)

Inky caps appear during spawn run on the mushroom beds or newly cased beds and outsides the manure piles during fermentation.

Symptoms

Inky caps are initially whitish and changes to bluish-black later on. Stalks and thin caps disintegrate into black slimy mass of spores. Infection comes from unpasteurized compost or casing soil or air.

Management

Use proper pasteurized compost or casing soil. Adjust pH to 8 and spray Bavistin (0.05 per cent). Eliminate ammonia from the compost before spawning. Rogue out these caps to avoid further spread.

10.1.9. Cinnamon Mould/Brown Mould

The appearance of this disease has been reported from J and K, Punjab and parts of H.P.

Symptoms

The mould first appears as large circular patches of mycelium on the compost/casing soil. Within few days spores are formed and the colour changes from white to light yellow or to golden brown. Fungus produces cup like fruit bodies. The spores are air borne.

Management

Casing soil should be properly sterilized. Spray Z-78 @ 0.2 per cent.

10.1.10. Lipstick Mould (*Sporendonema purpurescens*)

This disease has been reported from Punjab and H.P.

Symptoms

This disease appears in spawned compost as white crystal like colonies. On maturity spores colour changes from white to pink to cherry red and later dull orange. Primary source of inoculum is casing mixture and spent compost. In case of chicken manure in compost formulae, the litter is said to carry the lipstick fungus.

Management

Maintain the hygienic conditions to prevent the spread of disease. Proper sterilization of casing mixture should be done. Spent mushroom compost should be disposed off.

10.2. Bacterial Diseases

10.2.1. Bacterial Blotch

Bacterial blotch of the mushroom is also known as brown blotch and bacterial spot. Bacterial blotch disease reduces market value of crop because lesions develop on the surface of mushroom caps making the mushrooms unacceptable by consumers. It causes 5-10 per cent losses in yield.

Symptoms

It causes brown spots or blotches on the mushroom cap. These spots are irregular yellowish to dark brown and coalesce in the later stage. Mushrooms often become infected at a very early stage in their development. The enlargement of the spots on the cap surface is depend upon environmental conditions and is favoured by temperature (20°C) and presence of water film. The main source of infection is the casing soil. Infection spreads through water, flies and nematodes etc.

Management

Casing soil should be properly sterilized and ventilation should be quite adequate. Use of chlorinate water and application of streptomycin has been found effective in managing the disease. Spraying the casing soil with a mixture of *P. fluorescens* resulted in more than 80 per cent control.

10.2.2. Yellow Blotch (*Pseudomonas agarici*)

Heavy incidence of yellow blotch has been reported in oyster mushroom causing complete failure of crop.

Symptoms

At pinched stage it affects the total group of early fruit bodies or part of them. Fruit bodies turn yellow and stunted. When the relative humidity is more than 90 per cent, slimy appearance is a common symptom and fruit bodies starts rotting and foul smelling.

Figure 24: Mushroom Diseases.

Dry Bubble

Wet Bubble

Cobweb

Brown Plaster Mould

Green Mould

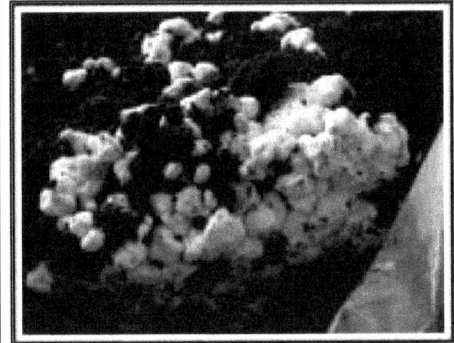

False Truffle

Contd...

Figure 24–*Contd...*

Olive Green Mould

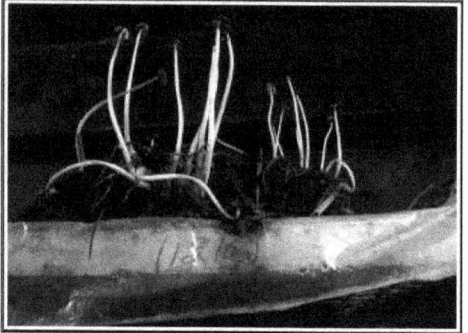

Ink Caps

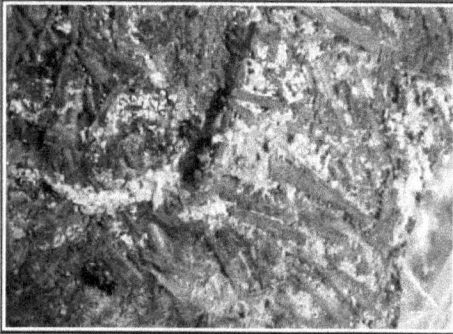

Lipstick Mould

Nematodes

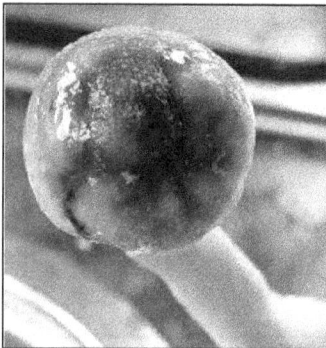

Bacterial Disease

Viral Disease

Contd...

Figure 24–*Contd...*

Physiological Disorders

Rose Comb

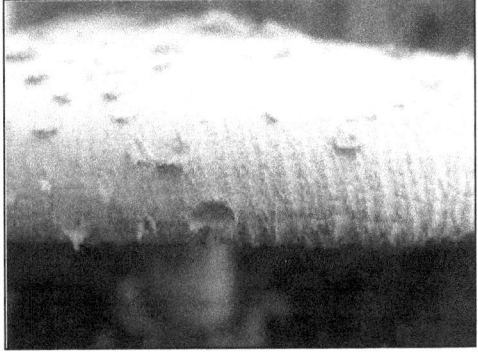

Hardening of Gills

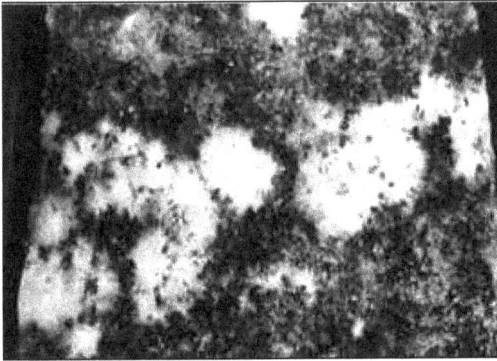

Stroma

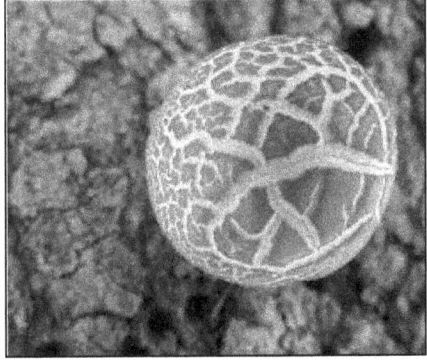

Cracking of Caps

Whereas, when relative humidity is <75 per cent the blotched fruit bodies give appearance of burnt ulcers. The spread is through water splashes, workers, tools and mushroom flies.

Management

Proper ventilation and watering coupled with monitoring of temperature in the mushroom unit helps in limiting the disease incidence.

Use chlorinated water spray and streptocyclin in the case of serious incidence.

10.3. Physiological Disorders

There are large numbers of abiotic agents which create unfavourable environment for proper growth of mushrooms, resulting in qualitative and quantitative loss.

1) Rose Comb

In this, cap is malformed and gills or lamella formed on its upper surface. On top of the cap of the mushroom appears a pink, wart like outgrowth composed of a mass of irregular and ill formed gills giving typical appearance of 'Cocks kalgi'. The abnormality is caused by the use of mineral oils or smoke of kerosene oil, coal or cigarette etc. in the mushroom house. Avoid smoke in the cropping room. If one is using kerosene oil stove for heating purposes then the fumes of the oil gas should not be circulated in mushroom house.

2) Hardening of Gills

It is due to frequent changes in temperature and humidity in cropping room. Gills become very hard. This can be managed by maintaining proper temperature and relative humidity at desired level.

3) Cracking of Caps

Due to high temperature or drought (*i.e.* above 20°C) and low humidity (below 80 per cent), surface of caps gives cracks appearance. This can be managed by reducing the temperature and maintaining 85 to 90 per cent relative humidity.

4) Long Stalked Mushrooms

In this, cap remains small and stalk elongated. This is due to excess of CO_2 in cropping room. This can be controlled by introducing fresh air immediately in mushroom house.

5) Opening of Caps

Due to the high temperature of the mushroom house and loss of moisture in the compost, premature opening of the caps occur at pin head stage. Reduce temperature and maintain proper moisture and relative humidity in the mushroom house.

6) Stroma and Sectoring

Aggregation of mushroom mycelium developed over the surface of spawned compost or on the casing soil. Stroma formed on the compost surface may be small and localized. Later the small patches can coalesce with each other covering larger areas of the surface.

A sector is portion of spawn which may be extra dense or extra fluffy and is always different from spawn. Few sectors may not affect yield but the presence of excessive stroma may reduce yield. Sectors are related to the genetic character of the spawn but may also be due to mishandled or exposed to harmful petroleum fumes/ detergent/chemicals. Patches of stroma should be removed from the compost and new growth of spawn may be normal.

10.4. Viral Diseases

Common Name

La France disease, brown disease, watery stipe, X disease or die back.

There are 3 types of viruses involved in die back symptoms. In India, virus and virus like diseases have been reported on button and oyster mushroom. Symptom expression depends upon the virus concentration, time of infection, strain of the spawn used and cultural and environmental conditions. The general symptoms observed are as below:

1. Mycelium does not permeate or hardly permeates the casing layer or disappears after normal spread. Mushrooms appear only in dense clusters, maturing too early.

2. Mycelium isolated from diseased sporophore on agar shows a slow and degenerated growth as compared with healthy mycelium.

3. Pin heads appear late and sometimes below the surface of casing layer. Soon after the sporophores appear above the casing layer, their pilei are already opened.

4. A specific musty smell can be perceived in growing room infected with the disease.

5. Diagnosis of mushrooms viruses is little difficult as same symptom can be induced by a variety of factors. The second difficulty with diagnosis is generally the low virus concentration. Among the tests ISEM and ELISA are presently being used widely.

Management of Mushroom Viruses

Hygienic measures are to be followed strictly. When virus-infected cultures are grown at 33°C for 2 weeks and hyphal tips then sub-cultured and returned to 25 °C, the newly obtained cultures are generally free from viruses.

For adopting suitable management strategies for mushroom viruses, one has to keep in mind that the disease is spread by viable mycelium and spores of diseased mushrooms; early infection is dangerous, especially an infection simultaneous with or shortly after spawning. Upto the time of casing, the compost and mycelium must be protected. Owing to the lack of useful resistance with the species, control of the disease is based largely on the use of hygienic practices directed at the elimination of diseased mycelium and basidiospores from the production

10.5. Nematodes

Nematodes are one of the major serious pest of mushroom. The multiplication rate has very fast (50-100 fold/week). During spawn run period the rate of multiplication is faster as compared to cropping period. If the compost is dried gradually, these nematodes survive in a state of anabiosis for upto two years but they die if the compost is dried rapidly. Some important nematode groups associated with mushrooms are given in Table 8.

Table 8: Nematodes Associated with Mushrooms

Feeding Group	Source of Contamination
Mycophages	Unpasteurized compost, casing soil, irrigation, water,etc.
Saprophages	Unpasteurized compost, casing soil, irrigation, water,etc.
Predatory	Compost and casing soil
Plant parasites	Casing soil

Symptoms

1. Whiteness of spawn run slowly changes to brown.
2. The compost surface sinks
3. Alternate high and poor yield in successive flushes.
4. Complete crop failure.
5. Decline in yields.
6. Mycelial growth is sparse, patchy and turns stingy.
7. Sporophores flushes are poor and delayed.

Management

1. Strict hygiene and sanitation
2. Composting yard must be cemented to prevent direct contact of compost with soil. The yard should be disinfected with 4 per cent formalin, 24 hrs before compost preparation.
3. Proper pasteurization of compost and casing material.
4. Use of clean irrigation water
5. All the instruments, walls, floors, gallaries should be disinfected with 4 per cent formalin.
6. All the rooms should be fly proof.
7. Foot dips must be installed in front of each cropping room.

11

Post Harvest Handling and Processing

Mushrooms are living microbial fruit bodies with high metabolic activities. On harvest, the actively growing fruiting bodies are excised from the substrate mycelium. Mainly four types of mushrooms *viz.*, white button (*Agaricus bisporus*),oyster (*Pleurotus spp.*),milky (*Calocybe indica*) and paddy straw mushroom (*Volvariella spp.*) are grown commercially in India. Of these, white button mushroom forms the major part of total production.

Fresh mushrooms are rich source of proteins with essential amino acids, minerals and vitamins. Phenols and enzymes such as oxidase, peroxidase and catalase are present at high level. Due to short life the product remains acceptable for few hours at ambient temperature of the tropics and sub tropics. Thus a knowledge of post harvest handling practices plays a significant role in enhancing the availability of quality mushrooms either in fresh or processed form to the consumers and at the same time ensuring remunerative prices to the producers.

11.1. Preservation and Processing of Mushrooms

Because of high moisture content (85-95 per cent) and delicate nature, mushroom are highly perishable and can't be stored for more than 24 hrs at ambient temperature. In the peak period of harvesting due to gluts in the market, preservation in to more stable products is essential.

11.1.1. Causes and Consequences of Mushroom Deterioration

a) Lack of protective covering leads to water loss, which leads to softness of body and texture deterioration.

b) High respiratory activity in the absence of new supply from substratum, the metabolism is disbalanced which leads to senescence and stalk elongation.

c) Absence of protective covering on the surface, disbalance in metabolism and lack of inherent antimicrobial mechanisms on harvest, make the mushroom prone to attack by associated microflora, which leads to discolourations, patches and slimy growth.

d) Presence of high levels of phenols/polyphenols and phenol oxidizing enzymes results in oxidation of colourless phenolic compounds in to highly coloured quinonic compounds and their complexes.

e) Richness in amino acids, proteins, vitamins and flavour compounds results into faster deterioration and development of off flavour.

f) Soft and delicate texture of mushroom fruiting bodies makes them highly susceptible to mechanical/physical damage.

In the peak period of harvesting the gluts in the market can be checked by adopting appropriate post harvest technologies. Development of appropriate storage and processing technology in order to extend their marketability and availability to consumer in fresh as well as processed form is of great significance. The post harvest technology includes:

11.2. Post Harvest Operations for Fresh Mushrooms

11.2.1. Harvesting

Harvesting of mushrooms at optimum stage of maturity is of great importance. Harvesting of under or over mature fruit bodies results in poor texture, flavour and immediate degradation.

11.2.2. Sorting

It is done to remove the undesirable fruiting bodies in the produce. In sorting various attributes like discolouration, blemishes and shape are taken in to consideration. This helps in getting premium price to the producer and delays the deterioration processes.

11.2.3. Pre-cooling

Although the mushrooms are grown in shades, these require to be pre-cooled immediately after harvest to their optimum storage temperature (5°C) to arrest the high respirative and deteriorative changes which are otherwise very fast at ambient room temperature.

11.2.4. Dips

To maintain whiteness, dipping of mushrooms in dilute solution of hydrogen peroxide (1:3) for half an hour and then steeping in 0.25 per cent citric acid solution containing 500 ppm sulphur dioxide has significant effect, however, it is a common practice to wash the produce in 0.025 to 0.05 per cent potassium metabisulphite solution to maintain the whiteness of the produce and to remove any adhering casing soil, compost or other substrates.

11.2.5. Packaging

Since mushrooms are very sensitive to desiccation and draught, selection of suitable package is very important. Mushrooms are usually packed in polypropylene bags of 200-500 g capacities. In 100 guage polyethylene bags 0.5 per cent ventilation is generally recommended for refrigerated storage whereas, for local markets non-perforated bags should be used.

11.2.6. Transportations

Mushroom being highly perishable and having a high rate of respiration, transport in refrigerated vans is recommended for long distance markets. For local market the pre-cooled mushroom packed in suitable packages should be transported in insulated ice containers.

11.2.7. Storage

During storage, metabolic activity of tissue continues where consumption of oxygen and production of carbon dioxide continues along with loss of moisture, which is dependent upon relative humidity of the surrounding. In an open atmosphere, relative concentrations of oxygen, carbon dioxide and water vapours remains steady and deterioration are affected through loss of moisture, weight and surface growth of contaminants. In case of closed system, oxygen depletion and carbon dioxide accumulation leads to anaerobiosis and shift in metabolism from respiratory to fermentative and putrefactive activity which is turn results in development of off flavour. Transpiration flow of water vapours leads to condensation and dew formation on containers internal surface. To avoid undesirable changes scientific consideration leads to two types of storage packaging systems based upon atmospheric controls: Controlled Atmosphere Storage (CAS) and Modified Atmosphere Packaging (MAP). In the CAS, a low temperature with more than 2 per cent oxygen and less than 10 per cent carbon dioxide is maintained by adding or deleting these constituents in storage cabins. In case of MAP, packaging materials and their thickness suitably selected for modifying the atmosphere through differential transfer rates of gases/gaseous constituents. A temperature of 5°C along with 85- 90 per cent relative humidity is generally recommended for the storage of mushrooms. At ambient temperature, the steeped mushrooms can be kept for 8-10 days.

11.3. Preservation Processes

In view of the highly perishable nature, the fresh mushrooms have to be preserved to extend their shelf-life to off season use. Different technologies for processing of mushroom are:

11.3.1. Low Temperature

Individual quick freezing (IQF) technology is employed to yield a product of excellent quality with longer shelf-life. Quick freezing is a process where temperature of the mushrooms passes thorough the zone of maximum crystal (0-3°C) in 30 min. or less. Freezing stops microbial activity and enzyme activity is retarded at freezing temperature. To control their activity mushrooms prior to freezing are required to be balanced. To obtain the best performance the storage temperature should be maintained at -35 °C to -40°C.

11.3.2. Dehydration

The main objective of drying is removal of free water to such a level that the biochemical and microbial activity is checked due to reduced water activity in the product. Blanching and sulphuring are important treatments given before drying. Pre- drying soaking of blanched mushroom in potassium metabisulphite (1 per cent) +citric acid (0.2 per cent)+salt (3.0 per cent) solutions for 16 hours followed by drying at 60±2°C for 8-10 hours yields best product with longer shelf life.

11.3.3. Canning

To destroy the microorganisms and to prevent the recontamination hermetic sealing is main principle of preservation by the use of high temperature. Generally, for canning purpose, small size buttons are required. Ascorbic acid, ethylene diamine tetra acetic acid (EDTA) and citric acid have been recommended as useful adjust for improvement of colour in canning of mushrooms.

11.3.4. Irradiation

This is the most effective method of preservation. Application of gamma radiation not only retards the deteriorative processes but also increases the shelf life of mushroom retaining the same quality up to 10-14 days if preserved at 5°C.

11.3.5. Chemical Preservation

There are several chemical additives which are non nutritive and are added in small quantities in foods to improve its appearance, flavour, texture and storage properties. These additives contribute substantially in the preservation. Mostly used additives are salt, sugar, acetic acid, vinegar, spices and oils etc. On the other hand, chemical preservatives maintain nutritional quality and enhance the keeping quality. Potassium metabisulphite is commonly used preservative in mushroom processing.

11.3.6. Lactic Acid Fermentation

Fermentation is a process of anaerobic or partial anaerobic oxidation of carbohydrates. During the process of fermentation sufficient quality of lactic acid is produced to prevent the product from further spoilage during storage.

12

Mushroom Recipes

12.1 Continental Mushroom Recipe

12.1.1. Appetizers

1. Antipasto Flat Mushrooms

Ingredients

- ☆ 4 large (about 125g each) flat Continental Mushrooms.
- ☆ 1/4 cup olive oil
- ☆ 1 ¹/² cups fresh breadcrumbs
- ☆ 3 large marinated artichokes, thinly sliced
- ☆ 60g small semi-dried tomatoes
- ☆ 60g roasted marinated capsicum chopped,
- ☆ 80g olives pitted and flesh chopped
- ☆ 60g feta cheese, crumbled,
- ☆ 2 tbs thyme leaves, grinded black pepper to taste.

Method of Preparation

Preheat a grill on medium-high. Brush both sides of mushrooms with 1 tbs oil. Place, stalk side up, onto grill tray and cook for 6 minutes or until tender. Meanwhile, heat remaining oil in a large frying pan over medium-high heat. Add breadcrumbs and cook, stirring frequently, for 5 or 6 minutes or until golden. Add artichokes, tomatoes, capsicum and olives to pan and cook, stirring occasionally, for 2 - 3 minutes or until warm. Remove from heat and stir in feta and thyme. Place mushrooms onto serving plates, top with antipasto mixture. Season with pepper and serve with crusty bread if desired.

Makes 4 servings

2. Crispy Herb Crumbed Mushrooms

Ingredients

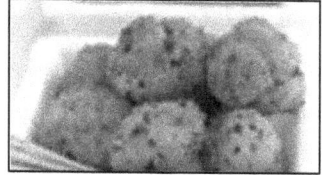

- ☆ 200g Continental Brown Mushrooms
- ☆ 3 eggs, lightly whisked
- ☆ 3 cups fresh breadcrumbs
- ☆ 1/4 cup finely chopped fresh parsley
- ☆ 2 tbs finely chopped chives
- ☆ salt and ground black pepper, to taste
- ☆ Canola oil for deep frying

Method of Preparation

Preheat oven to 150°C. Place whisked eggs in a shallow bowl. Combine breadcrumbs, parsley, chives, salt and pepper in another shallow bowl. Dip 1 mushroom into eggs and then coat well in the breadcrumb mixture, pressing breadcrumbs to secure. Place on a plate and repeat using remaining mushrooms. Heat oil in a medium saucepan or wok over medium heat. Deep-fry mushrooms in batches, turning occasionally, for 3 minutes or until golden.

Serve warm mushrooms with lemon wedges if desired.

Makes 4 servings

3. Make-Me-Drool Mushrooms

Ingredients

- ☆ 1/4 cup of oil, 2 medium sized onions
- ☆ 4-6 cloves of garlic
- ☆ 1 inch of fresh ginger
- ☆ 1 fresh green chilli
- ☆ 1 teaspoon of chilli powder
- ☆ 1 teaspoon of salt
- ☆ 1/2 teaspoon of turmeric powder
- ☆ 1 teaspoon of curry powder
- ☆ Handful of fresh coriander
- ☆ 1/4 teaspoon of garam masala
- ☆ 8oz Continental Sliced Mushrooms

Method of Preparation

Heat 6 tablespoon of oil for about 1 minute on high heat and then add 2 medium sized onions chopped, 4-6 cloves of garlic chopped, depending on how partial you are to garlic, 1 inch of ginger chopped

Heat onion mixture on medium heat for about 10 minutes, it is best not to let mixture stick to the bottom of the pan, add some water if the mixture starts sticking. After 10 minutes add: 1 small chilli cut lengthwise, 1 teaspoon of chilli powder, 1 teaspoon of salt, 1/2 teaspoon of turmeric powder, 1 teaspoon of curry powder.Stir in spices. Add a can of tomatoes or 4 peeled fresh tomatoes. Cover the pan and cook sauce on medium heat for about 10 minutes, after which you should have a nice thick sauce. Test the spices, if the sauce is too hot for your liking, or you would like to make it creamier then add about 3 tablespoons of natural yoghurt. Add mushrooms to sauce, turn the heat up and keep stirring. Turn down the heat after the mushrooms are nicely covered by sauce. Cover and simmer for 5 minutes. After 5 minutes add a handful of coriander leaves, chopped. Also add 1/4 teaspoon of *garam masala*. Serve with rice or *chapattis*.

Variation with Spinach

Add 1/2 pound of chopped baby spinach leaves instead of coriander and heat for a further 5 minutes before serving.

Variation with Sweet Peppers

Add 2 sweet peppers, cut into chunky pieces with the onion mixture.

Serves: 3-4

4. Mushroom and Zucchini Fritters

Ingredients

- ☆ 200g Continental Mushrooms thinly sliced
- ☆ 2 medium zucchini grated
- ☆ 2 tbs finely chopped fresh parsley
- ☆ 2 eggs separated
- ☆ salt and ground black pepper, to taste
- ☆ 2/3 cup self-raising flour
- ☆ 200g country-style natural yoghurt
- ☆ canola oil for frying
- ☆ 100g smoked salmon shredded

Method of Preparation

Combine mushrooms, zucchini, 1 tbs parsley, egg yolks and salt and pepper in a large bowl. Stir in flour and 2 tbs yoghurt, mix until well combined. Using electric hand-beaters, beat egg whites until soft peaks form. Fold one-third into mushroom mixture until just combined. Gently fold in remaining egg whites. Pour oil into a large frying pan to a depth of 1cm and heat over medium heat. Spoon 2 heaped tablespoons of fritter mixture into oil and slightly flatten. Cook fritters in batches for 2-3 minutes on each side or until golden and cooked through. Drain on paper towel. Stir remaining parsley through the remaining yoghurt. Serve fritters topped with yoghurt and smoked salmon.

Makes 12 serving

5. Mushroom Bruschetta

Ingredients

☆ 8 oz Continental Mushrooms

☆ 1/4 cup olive oil, 4 cloves garlic minced

☆ 3-4 large tomatoes, diced

☆ 1/3 cup Oregano, Salt and pepper

☆ 16 thick Italian bread slices.

Method of Preparation

Finely chop mushrooms. Heat 2 tbsp (25 ml) of the oil in skillet. Add mushrooms and cook, stirring, over medium heat, 2 minutes. Add garlic and cook another 2 minutes. Remove from heat; toss with tomatoes, oregano and salt and pepper to taste. Place bread slices on baking sheet. Drizzle with remaining oil; broil until golden brown, 1 to 2 minutes. Spoon mushroom mixture on top of each bread slice.

Makes 16 slices

12.1.2. Soups and Salads

1. Asian Style Chicken and Mushroom Soup

Ingredients

☆ 250g chicken tenderloins

☆ 2 tbs peanut oil

☆ 300g Continental Mushrooms sliced

☆ 1.25 litres chicken stock

☆ 1 bunch choy sum, chopped

☆ 250g fresh thin egg noodles

☆ 150g enoki mushrooms, trimmed

☆ 1 tsp sesame oil

Method of Preparation

Preheat a barbecue or chargrill on medium-high. Brush chicken with 1 tbs oil. Barbecue or grill chicken for 1-2 minutes on each side or until just cooked through. Cut chicken in half lengthways and keep warm. Heat remaining oil in a large saucepan on medium-high heat. Add mushrooms, cook 3-4 minutes or until tender. Stir in stock and choy sum stalks, reserving leaves for later. Bring to the boil, reduce heat to medium-low and simmer for 5 minutes. Add noodles, cook, stirring often, for 3 minutes. Add choy sum leaves, chicken and enoki mushrooms, cook for 1-2 minutes or until leaves have wilted. Spoon into serving bowls, drizzle with sesame oil and serve immediately.

Makes 4 serving

2. Mushroom and Chicken Caesar Salad

Ingredients
Dressing:

- ☆ 3 cloves garlic, minced
- ☆ 1/2 tsp salt
- ☆ 1/3 cup freshly grated Parmesan cheese
- ☆ 2 tbsp Dijon mustard, 1 tsp Worcestershire sauce
- ☆ 1/3 cup lemon juice
- ☆ 1 cup olive oil, 1 tsp sesame oil

Salad

- ☆ 4 boneless skinless chicken breasts (about 1 Ib/500 g)
- ☆ 3 tbsp olive oil
- ☆ 8oz Continental Mushrooms sliced
- ☆ 2 heads romaine lettuce, washed, dried and torn into pieces
- ☆ 1 cup croutons, 1/2 cup chopped, crisp bacon

Method of Preparation

Dressing: In blender or food processor, blend garlic, salt, cheese, mustard and Worcestershire sauce into a paste. Add lemon juice and blend. With blender running, slowly add oil and blend until combined.

Salad: Brush chicken with oil. Grill over medium heat. Cook 5 to 7 minutes per side or until chicken is golden brown on outside and no longer pink inside. Cut into thick diagonal slices; set aside.

In large shallow bowl, toss mushrooms and lettuce with dressing until well coated. Sprinkle with croutons and bacon. Arrange chicken decoratively on top.

Makes 4 servings

3. Mushroom Spilt Pea Soup

Ingredients

- ☆ 1 tbsp vegetable oil
- ☆ 1 onion, chopped
- ☆ 8 oz Continental Mushrooms sliced
- ☆ 1 large carrot, chopped
- ☆ 1 celery stalk, sliced
- ☆ 1 tsp each curry leaves and salt
- ☆ 4 cups water
- ☆ 1 cup split peas, 1 bay leaf, pepper to taste.

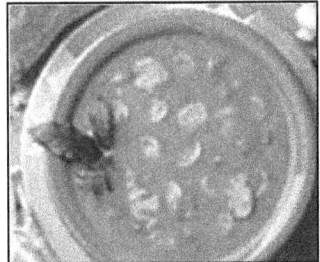

Method of Preparation

Heat oil in saucepan over medium heat. Add onion, mushrooms, carrot, celery and curry leaves. Cover and cook about 5 minutes or until onion is softened. Stir in salt, water, split peas and bay leaf. Bring to boil. Reduce heat to medium-low and cook covered about 40-60 minutes or until peas are very tender, stirring occasionally. Remove bay leaf. Season with pepper to taste. Soup freezes and reheats well.

Makes 4 servings

4. Mushroom, Prosciutto and Macadamia Salad

Ingredients
- ☆ 1/2 cup macadamia nut oil
- ☆ 3 tsp balsamic vinegar
- ☆ 400g Continental Mushrooms
- ☆ 125g thinly sliced prosciutto
- ☆ 100g macadamia nuts roughly chopped
- ☆ 100g baby spinach leaves, trimmed
- ☆ 250g cherry tomatoes, halved, salt and ground black pepper, to taste, crusty bread, to serve.

Method of Preparation

Whisk macadamia nut oil and vinegar in a large bowl until well combined. Add button mushrooms and stir to coat in oil mixture. Cover and stand for 15 minutes. Meanwhile, preheat a grill on medium-high. Lay the sliced prosciutto onto a grill tray and cook for 3 minutes on each side or until just golden. Remove to a plate. Cool and then break prosciutto into large pieces. Heat a large frying pan over medium-high heat. Add macadamia nuts and cook, stirring frequently, for 5 minutes or until golden. Remove to a plate. Add mushrooms to pan and cook, stirring occasionally, for 6 - 8 minutes or until just tender. Arrange spinach leaves, tomatoes, prosciutto, macadamia nuts and mushrooms onto serving plates. Season with salt and pepper and serve with bread.

Serves: 4

5. Spiced Mushroom, Corn and Noodle Soup

Ingredients
- ☆ 1 tbs olive oil
- ☆ 1 large red onion, cut into thin wedges
- ☆ 3 garlic cloves, crushed
- ☆ 2 small red chillies, finely chopped
- ☆ 1 litre chicken stock
- ☆ 400g Continental Mushrooms, thinly sliced
- ☆ 150g shiitake mushrooms
- ☆ 2 corn cobs, kernels removed
- ☆ 250g fresh Singapore noodles
- ☆ 2 tbs kecap manis
- ☆ 1-2 cup coriander leaves.

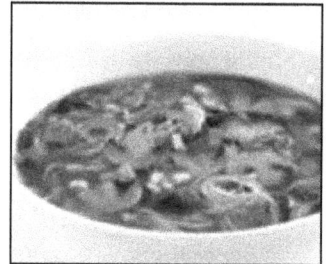

Method of Preparation

Heat oil in a large saucepan over medium-high heat. Add onion, garlic and chillies and cook, stirring frequently, for 4 minutes or until onion is soft. Stir in stock, mushrooms and corn. Bring to the boil, reduce heat to medium and simmer for 10 minutes. Add noodles and kecap manis and cook, stirring occasionally, for 3 minutes or until noodles are tender. Remove from heat, stir in coriander and serve immediately.

Serves: 4

12.1.3. Light and Hearty Meals

1. Heartland Steak and Grilled Mushrooms

Ingredients

- ☆ 1/2 cup unsalted butter
- ☆ 2 tablespoons Worcestershire sauce
- ☆ 1 teaspoon Dijon mustard
- ☆ 1/4 teaspoon salt
- ☆ 1/16 ground red pepper
- ☆ 1 lb. Continental White Mushrooms
- ☆ 4 beef filets (6-8 oz. each)

Method of Preparation

In small saucepan, melt butter; whisk in Worcestershire sauce, mustard, salt and red pepper. Simmer 1 minute; set aside. Thread mushrooms onto thin metal or wooden skewers that have been soaked in water; cover and refrigerate until ready to cook. grill steak over medium heat to desired doneness. Meanwhile, brush mushrooms generously with butter mixture and grill about 8-10 minutes, until lightly browned and tender crisp, brushing frequently with butter mixture. Arrange mushrooms over grilled steak. Reheat any remaining butter sauce and pour over all.

Makes 4 servings

2. Hot Mushroom Pasta

Ingredients

- ☆ 8 oz whole Continental Mushrooms
- ☆ 4 cloves garlic, minced
- ☆ 1/2 tsp red pepper flakes
- ☆ 1/4 cup olive oil
- ☆ 1 can (28 oz/796 ml) tomatoes, pureed
- ☆ 1 tsp pepper
- ☆ 3/41b long fusilli, cooked
- ☆ 1/2 cup grated Parmesan cheese
- ☆ 1/4 cup chopped fresh basil

Method of Preparation

Sauté mushrooms, garlic and pepper flakes in oil 4 minutes. Stir in tomatoes and pepper; cook 5 minutes. Pour sauce over pasta; top with cheese and basil.

Makes 4 servings

3. Mushroom and Almond Rice with Lamb

Ingredients

- ☆ $1^{1/2}$ cups long grain rice, rinsed
- ☆ 1/2 cup slivered almonds
- ☆ 1 tbs olive oil
- ☆ 5 green shallots thinly sliced
- ☆ 300g Continental Button Mushrooms, thinly sliced
- ☆ 1 orange, rind finely, grated and juiced
- ☆ 1/3 cup currants, salt and ground black pepper, to taste
- ☆ 12 lamb cutlets, trimmed

Method of Preparation

Cook rice in a large saucepan of boiling water for 10 - 12 minutes or until just cooked. Drain. Meanwhile, heat a large non-stick frying pan over medium high heat. Add almonds and cook, stirring constantly, for 3 minutes or until golden. Remove and set aside. Add oil, green shallots and mushrooms to frying pan and cook, stirring occasionally, for 2 - 3 minutes or until mushrooms are tender. Add orange rind and juice and cook for 30 seconds. Stir in rice, almonds and currants and toss gently over low heat until well combined. Season with salt and pepper. Keep warm. Preheat a grill or barbecue on medium-high. Season lamb cutlets with pepper. Cook for 3 - 4 minutes on each side or until cooked to your liking. Serve lamb cutlets with rice.

Makes 4 servings

4. Mushroom and Chicken Fajitas

Ingredients

- ☆ 1 cup red, green and yellow pepper strips
- ☆ 1 red onion, thinly sliced
- ☆ 3 tbsp vegetable oil
- ☆ 4 cups sliced Continental Mushrooms
- ☆ 1 Ib boneless, skinless chicken breasts, cut in strips
- ☆ 1/2 tsp each chili powder, cumin
- ☆ 8 tortillas, heated and Grated cheese.

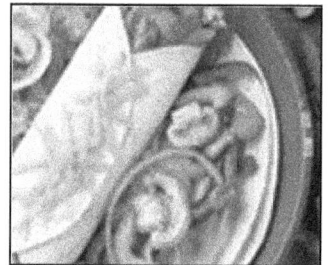

Method of Preparation

Sauté peppers and onion in 1 tbsp (15 ml) of the oil, 3 minutes. Add mushrooms; cook 5 minutes, stirring, set aside. Add remaining oil. Sprinkle seasonings over chicken; sauté, stirring 6 minutes, or until no longer pink. Toss with vegetables. Place in center of tortillas; roll up. Sprinkle with cheese.

Makes 8 fajitas

5. Mushroom and Chicken Tourtiere

Ingredients
- ☆ 1 lb ground chicken or turkey
- ☆ 1 cup chopped onion, 1 clove garlic, minced
- ☆ 1 tbsp oil
- ☆ 8oz Continental Mushrooms, sliced
- ☆ 1 cup chicken stock
- ☆ 3/4 cup fresh breadcrumbs
- ☆ 1 bay leaf, 1/2 tsp each thyme, marjoram and salt
- ☆ 1/4 tsp each ground cloves and black pepper
- ☆ 1 tbsp chopped fresh parsley
- ☆ 2 9-inch (23 cm) unbaked pie shells.

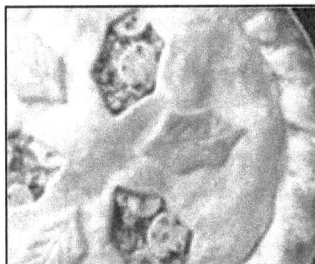

Method of Preparation
In large skillet, brown chicken with onion and garlic in oil until onion is limp. Stir in mushrooms. Cook and stir, 2 to 3 minutes. Add stock, breadcrumbs and spices. Cook, stirring occasionally, about 10 minutes. Discard bay leaf. Stir in parsley. Spoon mixture into one pie shell. Cut vents or decorative shapes in second pie shell. Place over filling. Crimp edges. Bake in 425°F (220°C) oven 15 minutes. Reduce heat to 350°F (180°C). Continue baking 10 to 15 minutes or until pastry is golden brown.

Makes 6 to 8 servings

6. Mushroom Focaccia Pizza

Ingredients
- ☆ 2 cloves garlic, crushed
- ☆ 15ml (1 tablesp) olive oil
- ☆ 1 large focaccia bread
- ☆ 30g ($1^{1/2}$ oz) sun-dried tomato paste
- ☆ 25g (1oz) butter
- ☆ 175g (6oz) Continental Mushrooms, thickly sliced
- ☆ 1 small pimento or red pepper de-seeded and cut into strips
- ☆ 4 cherry tomatoes, halved
- ☆ 6 pitted black olives
- ☆ 50g (2oz) goats cheese, sliced
- ☆ few fresh basil leaves, torn.

Method of Preparation
Preheat oven to 160°C, (325° F). Blend together the garlic and olive oil. Place the focaccia on baking tray and brush over the oil mixture. Spread focaccia with tomato paste. Melt butter in a frying pan and sauté mushrooms and pimento/pepper until soft. Spoon the mixture over bread then top with tomatoes, olives, goats cheese and

basil. Bake in the oven for 10 minutes or until cheese is bubbling. Garnish with some fresh basil.

Serves 4

7. Mushroom, Pork and Noodle Stirfry
Ingredients
- ☆ 300g pork fillet trimmed and thinly sliced
- ☆ 2 tbs peanut oil
- ☆ 2 tbs hoi sin sauce
- ☆ 2 tbs sweet chilli sauce
- ☆ 300g fresh Hokkien noodles
- ☆ 1 onion, cut into thin wedges
- ☆ 1 bunch baby bok choy chopped
- ☆ 300g Continental Mushrooms, halved.

Method of Preparation
Combine pork, 2 tsp oil, hoi sin and sweet chilli sauce in a medium bowl. Cover and refrigerate for 10 minutes. Place noodles into a heat-proof bowl, cover with boiling water, stand for 5 minutes. Drain, keep warm. Heat a wok over high heat. Add 2 tsp oil and half the pork, stir-fry for 2-3 minutes or until pork is tender. Remove and keep warm. Repeat using remaining pork. Heat remaining 2 tsp oil in wok, add onion, stir-fry for 1 minute or until tender. Add chopped bulbs of the bok choy, reserving leaves for later, and stir-fry for 1 minute. Add noodles, stir-fry for 2-3 minutes or until noodles are hot. Add mushrooms, pork and reserved bok choy leaves, stir-fry for 2-3 minutes or until pork is heated through. Serve immediately.

Serves 4

8. Mushroom Bacon and Herb Linguine
Ingredients
- ☆ 400g dried linguine
- ☆ 2 tbs olive oil, 1 onion, finely chopped
- ☆ 200g bacon rashers, rind removed and thinly sliced
- ☆ 500g Continental Mushrooms, thickly sliced
- ☆ 1/2 cup chicken stock
- ☆ 1/4 cup finely chopped continental parsley
- ☆ 1/4 cup thinly sliced chives, salt and ground black pepper, to taste.

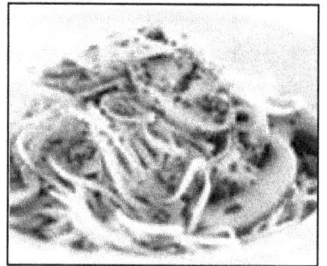

Method of Preparation
Cook linguine in a large saucepan of salted boiling water, following packet directions, until al dente. Meanwhile, heat oil in a large frying pan over medium-high heat. Add onion and cook for 3 minutes or until soft. Add bacon and cook, stirring

frequently, for 5 - 6 minutes or until bacon is crisp. Add mushrooms and cook, stirring occasionally, for 3 - 4 minutes or until mushrooms are tender. Stir in stock and cook for 1 minute or until hot. Drain linguine and return to the saucepan. Add mushroom mixture, parsley and chives. Toss gently over low heat for 2 minutes or until heated through. Season with salt and pepper. Serve immediately.

Serves 4

9. Mushroom, Beef and Vegetable Tacos
Ingredients
- ☆ 1 tbs olive oil
- ☆ 1 onion, finely chopped
- ☆ 400g lean beef mince
- ☆ 1 carrot, grated
- ☆ 1 zucchini, grated
- ☆ 250g Continental Mushrooms, thinly sliced
- ☆ 35g sachet taco seasoning mix
- ☆ 1/4 cup water
- ☆ 12 taco shells
- ☆ 4 lettuce leaves, shredded, to serve,
- ☆ 1 avocado, mashed, to serve.

Method of Preparation
Heat oil in a large non-stick frying pan over medium-high heat. Add onion and cook for 3 minutes or until soft. Add beef mince and cook, stirring often to break up mince, for 6 - 8 minutes or until browned. Add carrot, zucchini, mushrooms, taco seasoning mix and water. Cook, stirring occasionally, for 6 - 8 minutes or until vegetables are tender and liquid has evaporated. Meanwhile, preheat oven to 180°C. Place taco shells onto a baking tray and heat for 10 minutes or until hot. Place lettuce into base of taco shells, top with mince mixture. Serve immediately with avocado.

Makes 12 servings

10. Mushroom, Pork and Snow Pea Stir-Fry
Ingredients
- ☆ 400g pork fillet, thinly sliced
- ☆ 3 tsp grated fresh ginger
- ☆ 2 tbs kecap manis*
- ☆ 1 tbs sherry
- ☆ 2 tbs peanut oil
- ☆ 300g Continental Mushrooms, quartered
- ☆ 200g snow peas, trimmed

☆ 1/4 small red cabbage, shredded

☆ 2 tbs water, steamed rice, to serve.

Method of Preparation

Combine pork, ginger, kecap manis and sherry in a medium bowl. Cover and refrigerate for 10 minutes. Heat a wok over high heat. Add 3 tsp oil and heat until hot. Add half the pork and stir-fry for 2 minutes or until pork is tender. Remove and repeat using 3 tsp oil and remaining pork. Set pork aside. Add remaining 2 tsp oil, mushrooms, snow peas and red cabbage to wok and stir-fry for 1 minute. Add water, cover and cook for 1 minute or until vegetables are just tender. Remove cover, add pork and stir-fry for 1 - 2 minutes or until heated through.

Serve with steamed rice.

*KECAP MANIS IS THICK, SWEET SOY SAUCE AVAILABLE FROM SUPER MARKETS OR ASIAN FOOD STORES.

Serves 4

11. Portobellow Mushroom Stroganoff

Ingredients

☆ 2 tablespoons vegetable oil

☆ 6 ounces (or so) Continental Portobello Mushrooms, sliced about 1/4 to 1/2 inch thick

☆ 1 pound white mushrooms, sliced

☆ 4 cloves garlic, minced (about 2 teaspoons)

☆ 2 teaspoons fresh, minced rosemary leaves

☆ 1/4 teaspoon red pepper flakes

☆ 1 teaspoon salt

☆ 1 teaspoon dried tarragon or 1 tablespoon fresh

☆ 3 tablespoons (1/2 6-ounce can tomato paste)

☆ 8 ounces sour cream

Method of Preparation

Heat the oil in a wide skillet over medium-high heat. Add Portobello slices; cook 5 minutes, stirring occasionally. The oil in the skillet will dry up momentarily, but moisture from the mushrooms should seep back into the skillet. If it doesn't, turn the heat down. Add remaining mushrooms, garlic, rosemary, red pepper flakes, salt and tarragon. Cook 15 minutes, stirring occasionally, until mushrooms are quite dark and tender. Add 1/2 cup water and 3 tablespoons of tomato paste. Simmer a minute or two. Stir in sour cream and heat through, but do not boil. Serve over baked potatoes (or brown rice or noodles) along with kale or a salad.

Serves 4

12. Thai-Style Mushrooms with Noddles

Ingredients

- ☆ 500g hokkien noodles
- ☆ 1 tbs peanut oil
- ☆ 1 onion, cut into thin wedges
- ☆ 2 garlic cloves, crushed
- ☆ 1 bunch gai lum*,washed, leaves, separated and stems, chopped
- ☆ 2 tbs red curry paste
- ☆ 400g cup Continental Mushrooms, thickly sliced
- ☆ 1/2 cup coconut milk.

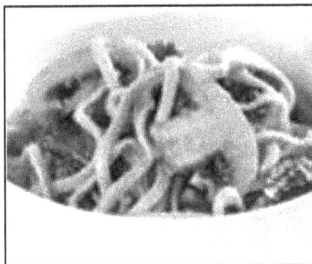

Method of Preparation

Place noodles into a large heat-resistant bowl and cover with boiling water. Stand for 2 - 3 minutes. Drain and separate noodles. Set aside. Heat a wok over high heat. Add oil and heat until hot. Add onion and garlic and stir-fry for 1 minute. Add stems of gai lum (reserving the leaves for later) and stir-fry for 1 minute or until bright green. Stir in curry paste and cook for 30 seconds. Add mushrooms, coconut milk, noodles and gai lum leaves and stir-fry for 2 minutes or until gai lum leaves have wilted and noodles are hot. Serve immediately.

*GAI LUM IS CHINESE BROCCOLI. BOK CHOY OR CHOY SUM MAY BE SUBSTITUTED.

Makes 4 servings

12.2. Mexican Mushroom Recipe

1. Mexican Chile and Mushroom Soup

Ingredients

- ☆ 3 large garlic cloves, left unpeeled
- ☆ 1 (1/2-inch-thick) slice large white onion
- ☆ 1 (3-inch-long) small dried ancho chile*
- ☆ (1/4 ounce)
- ☆ 1/2 cup water
- ☆ 3/4 teaspoon salt
- ☆ 1/2 teaspoon dried oregano (preferably Mexican), crumbled
- ☆ 2 tablespoons vegetable oil
- ☆ 10 ounces mushrooms, trimmed and thinly sliced
- ☆ 1 tablespoon tomato paste
- ☆ 3 1/2 cups low-sodium chicken broth (28 fluid ounces)

*Available at Latino markets, some specialty foods shops, some super markets, and Chile Today–Hot Tamale (800-468-7377).

Makes 4 first-course servings

Method of Preparation

Heat a dry 12-inch heavy skillet over moderate heat until hot, 3 to 5 minutes. Lightly smash garlic in skins with side of a large knife, then add to skillet along with onion slice and cook, turning over once or twice with tongs, until onion is well browned and garlic is slightly softened, about 8 minutes.

Meanwhile, discard stem, seeds, and veins from chile and tear chile into 4 pieces. Add chile to onion and garlic in skillet and toast, pressing flat with tongs and turning over occasionally, until chile turns a brighter red, about 1 minute.

Discard garlic skins and coarsely chop onion, then purée garlic, onion, and chile in a blender with water, salt, and oregano until smooth.

Heat oil in skillet over high heat until hot but not smoking, then sauté mushrooms, stirring occasionally, until golden and any liquid is evaporated, about 6 minutes. Reduce heat to moderate, then add tomato paste and cook, stirring, 1 minute. Add purée and cook, stirring, 3 minutes. Stir in broth and simmer 5 minutes.

2. Mushroom and Butternut Squash Empañadas (by chef Claire Archibald, cafe Azul, Portland)

Ingredients

For Empanada Filling

- ☆ 1 cup diced (1/4-inch) butternut squash
- ☆ 1/2 cup finely chopped white onion
- ☆ 6 small garlic cloves, minced
- ☆ 1/4 cup olive oil
- ☆ 2 (2- to 3-inch) fresh jalapeño chiles, seeds and ribs discarded and chiles finely chopped
- ☆ 1 pound fresh exotic mushrooms such as chanterelles, porcini, or hedgehogs (all one kind, not a mixture), trimmed and coarsely chopped
- ☆ 1/2 teaspoon salt, 1/3 cup chicken broth

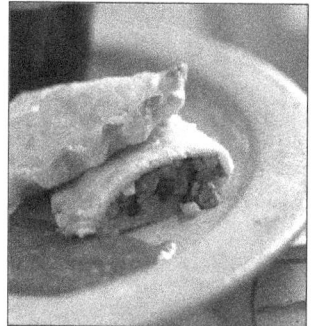

For Sauce

- ☆ 1 dried *pasilla de Oaxaca* chile*
- ☆ 3 garlic cloves, left unpeeled
- ☆ 1 pound fresh tomatillos, husks discarded and tomatillos rinsed and quartered
- ☆ 1/4 cup finely chopped white onion
- ☆ 1/4 cup water
- ☆ 1/2 teaspoon salt,

For Empanada Crust
- ☆ 1/3 Café Azul's pastry dough (1 pound)
- ☆ 1 large egg, lightly beaten with 1 tablespoon water
- ☆ 2 teaspoons coarse sea salt.

When buying the dried chile for this recipe, be aware that a *pasilla de Oaxaca* is not the same as a regular pasilla chile. The former is smoked and has a very distinct flavour.

Method of Preparation
Make Empanada Filling
Cook squash in a small saucepan of boiling salted water until just tender, about 2 minutes, then drain in a sieve. Cook onion and garlic in oil in a large heavy skillet over moderately low heat, stirring, until onion is softened, about 3 minutes. Add jalapeños and cook, stirring, 1 minute. Stir in mushrooms, salt, and broth and simmer, covered, until mushrooms are tender, 5 to 8 minutes. Simmer, uncovered, stirring occasionally, until liquid is evaporated, about 3 minutes, then stir in squash and salt to taste. Cool filling completely.

Make Sauce
Heat a dry griddle or heavy skillet (preferably cast-iron) over moderately low heat until hot, then toast *pasilla de Oaxaca* chile, pressing down with tongs, 15 to 20 seconds on each side. Halve chile lengthwise and discard stem, ribs, and seeds.

Heat griddle over moderately high heat until hot, then toast garlic until lightly blackened, 2 to 3 minutes on each side. Cool garlic slightly and peel.

Simmer tomatillos, onion, water, chile, garlic, and salt in a large saucepan, covered, until tomatillos are very tender, about 20 minutes, and cool slightly. Remove 1 chile half and reserve, then purée sauce in a blender until smooth (use caution when blending hot liquids), adding as much of reserved chile half as necessary to achieve desired spiciness. Return sauce to pan and season with salt.

Form and Bake Empanadas
Preheat oven to 400°F.

Divide dough into 8 equal pieces (2 ounces each) and form each into a disk. Roll out 1 piece on a lightly floured surface into a 6- to 7-inch round (1/8 inch thick). Spoon about 1/3 cup filling onto center and brush edge of pastry lightly with egg wash. Fold dough in half to form a half-moon, enclosing filling, and press edges together to seal. Crimp edge decoratively and transfer empanada with a spatula to a large baking sheet. Make 7 more empanadas in same manner.

Lightly brush empanadas all over with some of remaining egg wash and sprinkle each with 1/4 teaspoon sea salt. Bake in middle of oven until golden, 25 to 30 minutes.

While empanadas are baking, reheat sauce. Cut each empanada in half with a serrated knife and serve with about 3 tablespoons sauce spooned around it.

* Available at Latino markets and Kitchen/Market (888-468-4433).

Makes 8 first-course servings

2. Braised Mushroom Tacos (Recipe courtesy of fresh tastes)

Ingredients

- ☆ 2 small onions, sliced thin
- ☆ 3 tablespoons olive oil
- ☆ 12 grams (0.4 ounces) dried Guajillo chiles
- ☆ 12 grams (0.4 ounces) dried Ancho chiles
- ☆ 200 grams (7 ounces) canned whole tomatoes
- ☆ 2 tablespoons tahini
- ☆ 1/4 cup water
- ☆ 300 grams (10.5 ounces) Eryngi mushrooms (about 6 large ones)
- ☆ 140 grams (5 ounces) shiitake mushrooms (about 4 extra large ones)
- ☆ 2 large cloves of garlic, minced
- ☆ 2 teaspoons ground cumin
- ☆ 1/4 teaspoon ground cinnamon
- ☆ 1 teaspoon salt
- ☆ 12 corn tortillas, thinly shredded cabbage, cilantro, lime wedges.

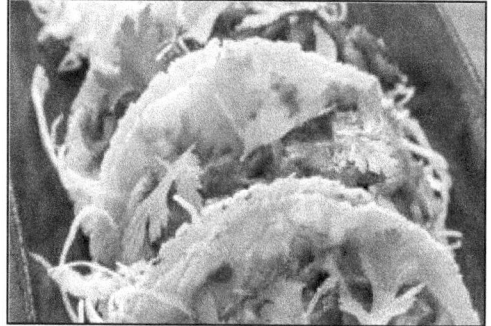

Method of Preparation

Place the onions and 1 tablespoon of olive oil in a sauté pan and fry over medium low heat until caramelized (20-30 minutes). If you have frozen caramelized onions you can skip this step. Boil a kettle of water. Use scissors to trim the tops off of the dried chilies then cut them from end to end so you can open them up. Remove all the seeds and membranes inside the chilies. Place the chilies in a bowl and cover with the boiling water. Let them soak for 15 minutes. While the chilies rehydrate, prepare the Eryngi mushrooms by cutting off the caps and using your fingers to pull apart the stems into thin shreds that look like shredded chicken. Slice up the caps along with the shiitake mushrooms. Drain the chilies and transfer them to a food processor along with the caramelized onions, tomatoes, tahini, and water. Add the remaining 2 tablespoons of olive oil to the sauté pan, and put over medium-high heat. Add the garlic, cumin and cinnamon. Fry until fragrant, then add the mushrooms and salt. Sauté until the mushrooms have browned. Strain the chili mixture through a sieve into the pan and turn the heat down to medium low. Simmer until the sauce is very thick. Adjust salt to taste. Serve with warm tortillas, shredded cabbage, cilantro and lime wedges.

4. Grilled Corn, Mushroom + Roasted Poblano Tacos with Chipotle Crema

Ingredients

- ✰ 5 large ears of sweet corn, grilled and cut off the cob,
- ✰ 2 large poblano peppers
- ✰ 2 cups sliced cremini mushrooms
- ✰ 1/2 teaspoon olive oil
- ✰ 1 ¹/² tablespoons unsalted butter
- ✰ 1/2 teaspoon salt
- ✰ 1/2 teaspoon pepper
- ✰ 1 large avocado, thinly sliced
- ✰ 4 ounces monterey jack cheese freshly grated
- ✰ 8-10 small corn tortillas, warmed

Method of Preparation

To roast peppers: remove core and seeds from peppers and slice into pieces. Lay on a baking sheet and preheat the broiler in your oven. Place under the broiler skin-side up until skins are completely charred and black – this took about 10 minutes but will depend on your oven. Just check every 2 minutes or so. Immediately remove peppers from oven and using kitchen tongs, quickly place them in a ziplock bag then seal it. Set aside for 20-30 minutes.

While peppers are "steaming" in the bag, heat a large skillet over medium low heat and add butter and olive oil. Toss in mushrooms, stirring to coat, then cover and let cook for 10-15 minutes, stirring occasionally. Remove from heat, season with salt and pepper and set side.

Assemble tacos by placing generous spoonfuls of corn, mushrooms and peppers in warm tortillas. Top with cheese, avocado slices and chipotle crema. Chipotle Crema [adapted from emeril] : 2 tablespoons sour cream, 3/4 cup half and half, 1 tablespoon adobo sauce (from a can of chipotles in adobo), juice of half a lime, 1/8 teaspoon salt.Combine all ingredients in a blender or food processor and blend until smooth. Emeril recommends letting his sit over night to develop culture)

5. Cheesy Mexican Mushroom Skillet

Ingredients

- ✰ 1 ¹/² teaspoons olive or vegetable oil
- ✰ 4oz uncooked vermicelli, broken into 1-inch pieces
- ✰ 1 medium onion, sliced (about 1 cup)
- ✰ 1 package (8 oz) sliced fresh mushrooms (about 3 cups)

☆ 1 can (14.5 oz) diced tomatoes

☆ undrained, 2 medium jalapeño chiles,

☆ seeded, finely chopped

☆ 1/2 cup water

☆ 2 teaspoons ground cumin

☆ 1/2 teaspoon salt

☆ 1 cup shredded Monterey Jack cheese (4 oz)

Method of Preparation

In 12-inch nonstick skillet, heat oil over high heat. Add vermicelli; cook about 2 minutes, stirring frequently, until golden brown. Reduce heat to medium. Stir in onion and mushrooms. Cook 2 minutes, stirring occasionally. Stir in tomatoes, chiles, water, cumin and salt. Reduce heat to medium-low; cover and cook 10 minutes, stirring occasionally. Remove from heat. Sprinkle with cheese. Cover; let stand about 2 minutes to melt cheese.

6. Easy Mexican Stuffed Mushrooms

Ingredients

☆ 10 Mushrooms (Cremini (brown) or white) - cleaned, steam removed and hollowed out

☆ 8 Cherry Tomatoes (quartered)

☆ 1/2 cup Sweet Corn

☆ 1/2 cup Kidney Beans

☆ 2 tablespoons fresh Cilantro (chopped)

☆ 3 Garlic Cloves (crushed)

☆ 1 teaspoon Cumin

☆ 1/4 teaspoon Salt

☆ 1/2 teaspoon Paprika

☆ Lime Juice (optional), Jalapeno (optional)

Method of Preparation

Clean mushrooms and remove stems. Carefully hollow out mushrooms with a small spoon, and dice what has been hollowed out. (stems will not be used). Combine diced mushroom with quartered tomatoes, corn, beans, cilantro, garlic, cumin, salt, and paprika. (Add diced jalapeno at this time if desired). Fill mixture into mushrooms using a spoon and push down with your fingers to ensure they are stuffed to their limit. Drizzle each mushroom with a dash of lime juice if desired.Place stuffed mushrooms on a foil pan and grill covered on medium-high heat for 6-8 minutes until mushrooms are tender (if using an oven see NOTES)

NOTES

If using an oven follow these directions: Heat oven to 435°F (225°C). Place stuffed mushrooms in a deep roasting pan. Cover roasting pan with tin foil and roast in oven (middle rack) for 12-15 minutes or until mushrooms are tender. The taste of raw mixture will be salty and overwhelmingly like garlic. After roasting or grilling the mushrooms the flavour will become milder.

7. Mexican Mussels with Sausage, Mushrooms, and Chiles

Ingredients

- ☆ 10 ounce longaniza sausage*, casing removed
- ☆ 3 tablespoons vegetable oil
- ☆ 2 cups chopped white onions
- ☆ 1 fresh poblano chile, stemmed, seeded, and chopped
- ☆ 1 red bell pepper, stemmed, seeded, and chopped
- ☆ 2 teaspoons kosher salt
- ☆ 1/2 teaspoon pepper
- ☆ 2 cups stemmed, sliced cremini mushrooms
- ☆ 4 large garlic cloves, minced
- ☆ 5 cups reduced-sodium chicken broth, divided
- ☆ 3 dried guajillo chiles*, stemmed and seeded, 2 pounds mussels, scrubbed

Method of Preparation

Cook sausage in an 8-qt. pot over medium heat, breaking up with a spoon, until cooked through and browned, about 10 minutes. With a slotted spoon, transfer sausage to a bowl lined with paper towels to absorb excess fat. Add oil, onions, poblano chile, bell pepper, salt, and pepper to pot. Cook, stirring often, until vegetables are softened and fragrant, about 8 minutes. Add mushrooms and garlic and cook 2 minutes longer. Add 1 qt. chicken broth and the cooked sausage. Bring mixture to a simmer, then reduce heat to low. Meanwhile, put guajillo chiles in a small saucepan with remaining 1 cup broth. Simmer until chiles have softened and turned the liquid reddish, 5 to 6 minutes. Transfer to a blender and whirl together. Strain chile mixture into pot with vegetables and sausage. Bring to a boil. Add mussels, cover pot, and turn off heat. Let sit until almost all the mussels have opened. Check after about 5 minutes; if they need more time, give a quick stir, re-cover, and let sit another 1 to 2 minutes. Pour mixture into a large serving bowl. * Longaniza sausage (Amelia likes El Mexicano brand) and guajillo chiles are found in most Latino markets.

12.3. Italian Mushroom Recipe

1. Italian Mushrooms

Ingredients

☆ 1 pound medium fresh mushrooms

☆ 1 large onion, sliced

☆ 1/2 cup butter, melted

☆ 1 envelope Italian salad dressing mix.

Method of Preparation

In a 3-qt. slow cooker, layer mushrooms and onion. Combine butter and salad dressing mix; pour over vegetables. Cover and cook on low for 4-5 hours or until vegetables are tender. Serve with a slotted spoon.

Makes 6 servings.

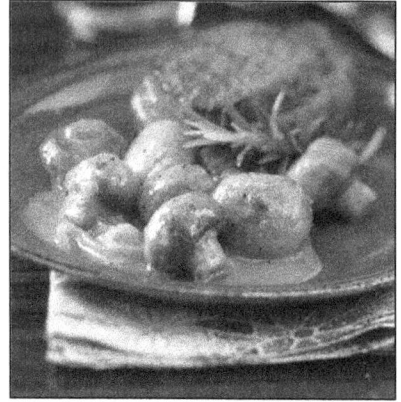

2. Italian Style Baked cheesy Mushrooms (With smoky bacon bits and crunchy croutons)

Ingredients

☆ 8 thin slices good-quality ciabatta bread, or a few carta di musica (music bread) broken up into rough pieces, extra virgin olive oil

☆ sea salt

☆ freshly ground black pepper

☆ 2 cloves garlic

☆ peeled and finely sliced

☆ 8 rashers higher-welfare smoked streaky bacon, roughly chopped

☆ a few sprigs fresh thyme, leaves picked

☆ 4 good handfuls mixed wild mushrooms (such as chanterelles, girolles blewits and morels)

☆ 1 dried chilli

☆ 125 g scamorza or buffalo mozzarella ball

☆ 2 handfuls rocket, washed and spun dry

☆ 2 heads dandelion, washed and spun dry

☆ 2 handfuls watercress, washed and spun dry

☆ 1/2 lemon

Method of Preparation

Preheat the oven to 180°C/350°F/gas 4. Heat a griddle pan and toast the ciabatta slices until they have lovely dark griddle marks. Tear up the toasted bread into chunks or, if using carta di musica, break up into pieces and toss with a drizzle of olive oil and a good pinch of salt and pepper. Get yourself a medium-sized earthenware dish, drizzle it with a little olive oil then add the seasoned, toasted bread.In the same bowl you used for the bread, add the sliced garlic, chopped bacon and thyme leaves, then tear over the mushrooms, leaving any little ones whole. Crumble in the dried chilli, then use your hands to toss everything together and get all those wonderful flavours going. Sprinkle this mixture fairly evenly on top of the bread, then tear over big pieces of the scamorza. Give the whole lot one final drizzle of olive oil, and put it in the oven to bake for about 30 minutes, until the mushrooms are beginning to crisp up and the cheese is melted, bubbling and starting to brown. While that bakes, wash and dry the salad leaves, and make a simple dressing by squeezing the juice of half a lemon into a jam jar. Top with twice as much extra virgin olive oil, and add a good pinch of salt and pepper. Put the lid on and give it a good shake, then put it to one side. Once the mushrooms are ready, drizzle some of your jam-jar dressing over the salad (any leftover dressing will keep happily in the fridge for a few days). Serve your beautiful, baked mushrooms and salad in the middle of the table and let everyone dig in.

3. Italian Sausage and Wild Mushroom Risotto

Ingredients

- ☆ 2 tablespoons olive oil
- ☆ 1 pound Italian sweet sausage, casings, removed, crumbled into 1/2-inch pieces
- ☆ 8 ounces portobello mushrooms, stemmed, dark gills scraped out, caps diced
- ☆ 10 ounces fresh shiitake mushrooms, stemmed, diced
- ☆ 1 teaspoon chopped fresh thyme
- ☆ 1 teaspoon chopped fresh oregano
- ☆ 1 1/2 cups Madeira
- ☆ 6 cups chicken stock or canned low-salt chicken broth
- ☆ 1/2 cup (1 stick) butter
- ☆ 1 large onion, chopped
- ☆ 4 garlic cloves, minced
- ☆ 2 cups arborio rice or other medium-grain rice (about 13 ounces)
- ☆ 1 cup freshly grated Asiago cheese*

At the restaurant, this dish is served as a starter. It would also make a great main course for four.

Method of Preparation

Heat oil in large nonstick skillet over medium-high heat. Add sausage and sauté until beginning to brown, about 3 minutes. Add all mushrooms, thyme, and oregano and sauté until mushrooms are tender, about 10 minutes. Add 1/2 cup Madeira; boil until almost absorbed, about 1 minute. Set aside. Bring stock to simmer in large saucepan; remove from heat and cover to keep hot. Melt butter in heavy large pot over medium-high heat. Add onion and garlic and sauté until onion is translucent, about 5 minutes. Add rice; stir 2 minutes. Add remaining 1 cup Madeira; simmer until absorbed, about 2 minutes. Add 1 cup hot stock; simmer until almost absorbed, stirring often, about 3 minutes. Continue to cook until rice is just tender and mixture is creamy, adding more stock by cupfuls, stirring often and allowing most stock to be absorbed before adding more, about 25 minutes. Stir in sausage mixture. Season to taste with salt and pepper. Transfer to serving bowl. Pass cheese separately.

4. Mushrooms and Spinach Italian Style (Recipe by Salvatore)

Ingredients

- ☆ 4 tablespoons olive oil
- ☆ 1 small onion, chopped
- ☆ 2 cloves garlic, chopped
- ☆ 14 ounces fresh mushrooms, sliced
- ☆ 10 ounces clean fresh spinach, roughly chopped
- ☆ 2 tablespoons balsamic vinegar
- ☆ 1/2 cup white wine
- ☆ Salt and freshly ground black pepper to taste, chopped fresh parsley, for garnish.

Method of Preparation

Heat the olive oil in a large skillet over medium-high heat. Saute onion and garlic in the oil until they start to become tender. Add the mushrooms, and fry until they begin to shrink, about 3 to 4 minutes. Toss in the spinach, and fry, stirring constantly for a few minutes, or until spinach is wilted. Add the vinegar, stirring constantly until it is absorbed, then stir in the white wine. Reduce heat to low, and simmer until the wine has almost completely absorbed. Season with salt and pepper to taste, and sprinkle with fresh parsley. Serve hot."This recipe is a typical recipe of Southern Italy, specifically Apulia. Spinach and mushrooms are sauteed with onion, garlic, balsamic vinegar, and white wine."

For 4 servings

5. Italian Stuffed Mushrooms

Ingredients

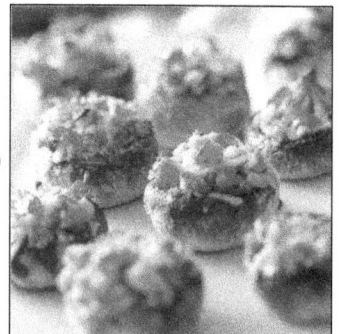

- ☆ Medium whole fresh mushrooms (1 pound)
- ☆ 2 tablespoons butter or margarine
- ☆ 1/4 cup chopped green onion (3 medium)
- ☆ 1/4 cup chopped red bell pepper
- ☆ 1 1/2 cups soft bread crumbs

- ✰ 2 teaspoons Italian seasoning
- ✰ 1/4 teaspoon salt
- ✰ 1/4 teaspoon pepper
- ✰ 1 tablespoon butter or margarine Grated Parmesan cheese, if desired

Method of Preparation

Heat oven to 350°F. Twist mushroom stems to remove from mushroom caps. Finely chop enough stems to measure 1/3 cup. Reserve mushroom caps. Melt 2 tablespoons butter in 10-inch skillet over medium-high heat. Cook chopped mushroom stems, onions and bell pepper in butter about 3 minutes, stirring frequently, until onions are softened; remove from heat. Stir in bread crumbs, Italian seasoning, salt and pepper. Fill mushroom caps with bread crumb mixture. Melt 1 tablespoon butter in rectangular pan, 13x9x2 inches, in oven. Place mushrooms, filled sides up, in pan. Sprinkle with cheese. Bake 15 minutes. Set oven control to Broil. Broil mushrooms with tops 3 to 4 inches from heat about 2 minutes or until tops are light brown. Serve hot.

12.4. Chinese Mushroom Recipe

1. Mu Shu Pork (By Rhonda Parkinson)

Ingredients

- ✰ 1/2 pound pork tenderloin Marinade
- ✰ 2 teaspoons dark soy sauce
- ✰ 2 teaspoons Shao-Hsing rice wine or dry sherry
- ✰ 1 teaspoon sugar
- ✰ 2 teaspoons cornstarch

Sauce

- ✰ 3 tablespoons water
- ✰ 3 tablespoons low-sodium chicken broth
- ✰ 1 tablespoon soy sauce
- ✰ 2 teaspoons Shao-Hsing wine or dry sherry
- ✰ 1/2 teaspoon salt
- ✰ 1/2 teaspoon sesame oil
- ✰ 1 teaspoon cornstarch

Other

- ✰ 4 dried black mushrooms
- ✰ 4 tablespoons dried cloud ears or wood ears
- ✰ 2 tablespoons dried lily buds
- ✰ 1/2 cup canned bamboo shoots
- ✰ 2 green onions

☆ 2 slices ginger, 2 eggs

☆ 1/2 teaspoon salt

☆ 6 tablespoons oil for stir-frying, or as needed, 1 teaspoon sesame oil (Prep Time: 30 minutes, Cook Time: 10 minutes)

Method of Preparation

Cut the pork into thin strips. Add the marinade ingredients, adding the cornstarch last. Marinate the pork for 30 minutes. Mix the sauce ingredients, whisking in the cornstarch last. Set aside. Place the dried mushrooms, cloud ears or wood ears, and lily buds in separate bowls and soak for approximately 30 minutes. Squeeze out any excess water. Remove the stems from the black mushrooms and the hard tips from the lily buds. Cut into thin strips. Rinse the bamboo shoots under warm running water to remove any tinny taste. Drain and cut into thin strips. Clean the green onion and cut into thirds. Cut the ginger into thin strips. Lightly beat the eggs with the salt. Heat the wok over medium-high to high heat. Add 2 tablespoons oil, swirling along the sides. When the oil is hot, add beaten eggs and scramble until they are quite firm. Remove the eggs from the wok. Clean out the wok. Add 2 tablespoons oil. When the oil is hot, add the ginger and stir-fry until aromatic (about 30 seconds). Stir-fry until the pork changes colour and is nearly cooked through. Remove from the wok. Add 2 tablespoons oil. When the oil is hot, add the wood ears, lily buds, and mushrooms. Stir-fry for about 1 minute. Push up to the sides. Give the sauce a quick re-stir and add in the middle of the wok, stirring quickly to thicken. Add the other pork and scrambled egg back into the pan. Stir in the green onion. Mix everything together. Taste and add extra seasoning if desired. Remove from the heat and stir in the sesame oil.

To Serve: serve with mandarin crepes and hoisin sauce. Place pancake on a plate and brush with hoisin sauce. Add meat mixture and roll up the pancake. *If desired, instead of stir-frying the green onion you can add it to the meat mixture at this point.

2. Shrimp with Chinese Greens Stir Fry (By Rhonda Parkinson)

Ingredients

☆ 1/2 - 3/4 pound shrimp

☆ 1 tablespoon Chinese rice wine or dry sherry

☆ 1/2 teaspoon salt, 1 tablespoon cornstarch

☆ 1/2 pound Chinese greens (bok choy)

☆ 4 ounces fresh mushrooms, or 6 Chinese dried mushrooms (also called dried Shiitake mushrooms)

☆ 2 tablespoons vegetable or peanut oil for stir-frying, or as needed

☆ 2 thin slices ginger

☆ 1/4 teaspoon salt,

☆ 1/4 cup chicken broth sodium-reduced if possible

☆ 1/2 teaspoon sugar

☆ 1 tablespoon light soy sauce

☆ black pepper, to taste, 1 teaspoon cornstarch mixed with 2 teaspoons water

Method of Preparation

If using frozen shrimp, defrost in the refrigerator. Rinse the shrimp under cold running water and pat dry with paper towels. Place the shrimp in a bowl and add the rice wine or sherry, 1/2 teaspoon salt and cornstarch, stirring in one direction (this is to make sure the marinade spreads evenly). Chop the bok choy stalks diagonally and the leaves across into 1 inch pieces. Wipe the mushrooms with a cloth or soft brush and cut into thin slices. If using Chinese dried mushrooms, soak in hot water for 20 minutes to soften. Drain the softened mushrooms, remove the stems and cut into quarters. Preheat the wok and add 2 tablespoons oil. When the oil is hot, add the ginger. Stir-fry for about 30 seconds, until aromatic, then add the prawns. Stir-fry until they turn pink. Remove the cooked shrimp from the pan. Add a bit more oil if needed so that there is about 1 1/2 tablespoons oil in the wok. Add the bok choy, mushrooms and 1/4 teaspoon salt. Stir-fry for 1 minute (Note: add a small amount of water or rice wine if the vegetables are a bit dry). Add the chicken broth, cover and cook for 2 more minutes. Add the shrimp back into the pan. Add the sugar, soy sauce, and pepper. Give the cornstarch/water mixture a quick restir and add in the middle, stirring to thicken. Cook, stirring for another minute and serve hot.

3. Shu Mai (Siu Mai) Dumplings (By Rhonda Parkinson)

Ingredients

☆ 1/2 lb ground pork

☆ 2 dried chinese mushrooms, soaked, then trimmed and chopped fine

☆ 1 tablespoon chopped fresh green ginger

☆ 2 green onions chopped fine

☆ 1 tablespoon soy sauce

☆ 1 teaspoon rice wine

☆ 1 teaspoon sesame oil

☆ 2 tablespoons potato starch, dash of salt, or to taste

☆ 24 gyoza wrappers

Prep Time: 30 minutes

(Cook Time: 7 minutes, Total Time: 37 minutes)

Method of Preparation

Combine all ingredients except for the wrapper, mixing well. Place about a tablespoon of filling on each wrapper, and gather up the sides to form ripples, leaving the center open. Wack the bottom of the dumpling on the counter so that it will stand up. When all the shu mai are filled, steam in a bamboo steamer for about 5-6 minutes, until cooked. Served with sweet and sour sauce or other dipping sauces.

4. Soy Sauce Chicken with Shiitake Mushrooms (By Rhonda Parkinson)

Ingredients

- ☆ Marinade: 1/3 cup soy sauce
- ☆ 1/3 cup water
- ☆ 3 tablespoons dark soy sauce
- ☆ 4 tablespoons brown sugar
- ☆ 2½ tablespoons Chinese rice wine or dry sherry
- ☆ 1½ green onions, washed and finely chopped
- ☆ 3 slices ginger
- ☆ 2 crushed cloves garlic

Other

4 chicken thighs, skin removed, 6 dried shiitake mushrooms 2 tablespoons olive oil or other vegetable oil, 1 small onion, peeled and chopped, Extra salt, and freshly ground black or white pepper, to taste

(Prep Time: 30 minutes, Cook Time: 45 minutes)

Method of Preparation

Mix together the marinade ingredients. Make 2 or 3 diagonal cuts on each side of the chicken thighs. Place the thighs in a large resealable plastic bag and add the marinade. (Use 2 bags and split the marinade in half if necessary). Marinate the chicken in the refrigerator for 4 hours or longer, moving the bag occasionally to make sure all the thighs are coated in the marinade. Remove the chicken and reserve the marinade. While the chicken is marinating, soften the dried shiitake mushrooms by soaking in hot water for about 30 minutes. Squeeze out any excess water. Cut off the stems and cut the caps in half. In a large saucepan, heat the olive oil over medium heat. Add the chicken thighs and cook for 5 to 6 minutes until browned, turning over once. Add the onion and saute until softened (about 5 minutes). At the same time you are browning the chicken and cooking the onions, bring the reserved marinade to boil in a small saucepan. Boil the marinade for 5 minutes. Add the marinade and the dried mushrooms to the chicken and onions. Season with salt and freshly ground black or white pepper if desired. Simmer, uncovered, on low heat for 30 minutes, or until the chicken is cooked through, adding water if needed. Serve soy sauce chicken hot over steamed rice.

5. Cantonese Steamed Chicken (By Rhonda Parkinson)

Ingredients

- ☆ 4 - 6 medium sized Chinese dried mushrooms (reserve 1 tablespoon soaking liquid)
- ☆ 1½ pounds assorted chicken pieces Bone
- ☆ 1/4 teaspoon salt Pepper, to taste
- ☆ 1½ tablespoons soy sauce (4½ teaspoons)
- ☆ 1 tablespoon Chinese rice wine or dry sherry
- ☆ 1 teaspoon sugar
- ☆ 1 teaspoon sesame oil
- ☆ 1½ tablespoons cornstarch
- ☆ 2 slices ginger, shredded (about 1 tablespoon)
- ☆ 1 green onion (spring onion, scallion), diced

(Prep Time: 15 minutes, Cook Time: 30 minutes)

Method of Preparation

Soak the dried mushrooms in a bowl of warm water for 20 minutes, or until they have softened. Squeeze out the excess water, cut the stems off the mushrooms and thinly slice. Reserve 1 tablespoon of the mushroom soaking liquid. Use a heavy cleaver to chop the chicken through the bone into bite-sized pieces. Place in a heatproof bowl and add the salt, pepper, soy sauce, rice wine or sherry, sugar, sesame oil, reserved mushroom liquid and cornstarch. Allow to marinate while bringing water to a boil for steaming. Place the bowl on a rack in a pot for steaming or in a steamer such as a bamboo steamer in a wok. Place the chicken pieces in the middle of the plate and surround with the mushrooms. Sprinkle the shredded ginger and green onion over top. Steam the chicken over boiling water for 15 - 20 minutes, making sure it is thoroughly cooked. Serve over rice.

6. Chinese Stir-fry Beef Recipe - Stir-fry Beef with Oyster Sauce (By Rhonda Parkinson)

Ingredients

- ☆ 1 pound beef steak, round or flank

Marinade

- ☆ 1 1/2 tablespoons dark soy sauce

☆ 1 tablespoon Chinese rice wine or dry sherry

☆ 2 teaspoons cornstarch

☆ 1 ¹/² tablespoons water

☆ 1 tablespoon vegetable oil.

Other

☆ 1/4-inch slice ginger, chopped

☆ 3/4 cup sliced mushrooms

☆ 1 small carrot, peeled and cut diagonally into thin slices

☆ 2 ¹/² tablespoons oyster sauce

☆ 1/2 teaspoon granulated or soft brown sugar

☆ 1/4 cup water or low-sodium chicken broth,

☆ 4 tablespoons oil for stir-frying or as needed, salt and pepper, to taste

(Prep Time: 15 minutes, Cook Time: 7 minutes, Total Time: 22 minutes)

Method of Preparation

Cut the beef across the grain into thin slices, 2 to 3-inches long and 1/4-inch thick. Add the marinade ingredients in the order given. Marinate the beef for 15 minutes. In a small bowl, mix the water or chicken broth, sugar and oyster sauce together and set aside. Heat 2 tablespoons oil over medium-high to high heat in a preheated wok. Add the ginger and stir-fry quickly until aromatic. Add the beef. Brown for a minute, then stir-fry until it changes colour and is nearly cooked through. (Cook the beef in two batches if needed). Remove the beef slices from the wok and drain. Clean out the wok. Heat 2 tablespoons oil in the wok. Add the sliced carrot, stir-fry briefly then add the mushrooms. Push the vegetables to the side and add the sauce in the middle. Let come to a boil, then add the beef back into the pan. Mix everything together - taste and adjust seasoning if desired. Served hot with steamed rice.

7. Stir-fried Mushrooms with Oyster Sauce

Ingredients

☆ 15 - 20 dried black mushrooms

☆ 2 cloves garlic, 2 slices ginger

☆ 2 tablespoons vegetarian oyster sauce

☆ 2 tablespoon dark soy sauce

☆ 2 teaspoons Chinese rice wine or dry sherry

☆ 1 teaspoon granulated sugar

☆ 1/3 cup reserved mushroom soaking liquid or chicken broth* (See Note Below) a few drops sesame oil, 1 green onion chopped

(Prep Time: 20 minutes, Cook Time: 7 minutes, Total Time: 27 minutes)

Method of Preparation

Reconstitute the mushrooms by soaking in warm water for 20 minutes or until they have softened. Squeeze to remove any excess liquid and slice. Reserve 1/3 cup of the soaking liquid (strain to remove any grit if necessary). Mince the garlic and ginger. Cut the green onion on the diagonal into 1-inch pieces. In a small bowl, combine the oyster sauce, dark soy sauce, Chinese rice wine or sherry, sugar and the reserved mushroom soaking liquid. Set aside. Heat a wok over medium-high to high heat. Add 2 tablespoons oil to the heated wok. When the oil is hot, add the garlic and ginger. Stir-fry briefly until aromatic. Add the dried mushrooms. Stir-fry for about 1 minute, then add the sauce. Bring to a boil and stir-fry until the mushrooms are coated with the sauce. Stir in the green onion and the sesame oil. Serve immediately.

*You can use chicken broth in place of the mushroom soaking broth if you're not worried about keeping the dish vegetarian.

8. Stir-fry Fish Cubes (By Rhonda Parkinson)

Ingredients
☆ 1 pound whitefish fillets

Marinade
☆ 1 egg white, pinch of salt
☆ 1 teaspoon cornstarch

Sauce
☆ 1/4 cup mushroom soaking liquid
☆ 1 teaspoon granulated or brown sugar
☆ few drops sesame oil
☆ Salt and pepper, as desired

Other
☆ 1 red bell pepper
☆ 4 large Chinese dried mushrooms
☆ 2 carrots, 2 slices ginger
☆ 1 green onion, chopped
☆ 1 teaspoon cornstarch mixed in 4 teaspoons water

(Prep Time: 0 minutes, Cook Time: 7 minutes, Total Time: 7 minutes)

Method of Preparation

Rinse the fish fillets and pat dry with paper towels. Cut the fish into bite-sized cubes. Mix the fish with the marinade and marinate for 25 minutes. While the fish is marinating, prepare the sauce and vegetables. Cut the bell pepper in half, remove the stem and seeds, and cut into cubes. Reconstitute the dried black mushrooms by soaking in warm water until softened (20 - 30 minutes). Remove the mushrooms, squeeze dry, and slice. Strain the mushroom soaking broth and save 1/4 cup. Peel and slice the carrots on the diagonal. In a small bowl, combine the strained mushroom soaking liquid, sugar, sesame oil and salt and pepper. Set aside. In another small bowl, dissolve the cornstarch in the water and set aside. Heat the wok and add 1 cup oil. When the oil is hot, poach the fish in the hot oil for 1 minute until golden. Remove and drain on paper towels. Drain all but 2 tablespoons from the wok. Add the ginger and stir-fry until fragrant. Add the carrots. Stir-fry briefly, then add the red bell pepper cubes. Add the mushrooms. Add the sauce into the middle of the wok. Turn up the heat and add the cornstarch/water mixture, stirring quickly to thicken. Add the fish back into the pan. Stir in the green onion. Add a bit of sesame oil if desired. Mix everything through and serve hot.

9. Buddha's Delight (By Rhonda Parkinson)

Ingredients

- ☆ 4 dried Shiitake or Chinese black mushrooms
- ☆ 1/2 cup dried lily buds
- ☆ 4 dried bean curd sticks
- ☆ 8 ounces bamboo shoots
- ☆ 6 fresh water chestnuts
- ☆ 2 large carrots,
- ☆ 1 cup shredded Napa cabbage
- ☆ 4 ounces snow peas
- ☆ 1/4 cup canned gingko nuts
- ☆ 1 knuckle of ginger, crushed

Sauce

- ☆ 4 tablespoons reserved mushroom soaking liquid or vegetable stock
- ☆ 1 tablespoon Chinese rice wine or dry sherry
- ☆ 1 tablespoon dark soy sauce
- ☆ 1 teaspoon sugar
- ☆ 1/2 teaspoon sesame oil

Other: Vegetable or peanut oil for stir-frying, as needed Salt, Accent or MSG to taste

(Prep Time: 30 minutes, Cook Time: 15 minutes, Total Time: 45 minutes)

Method of Preparation

In separate bowls, soak the mushrooms, dried lily buds, and dried bean curd sticks in hot water for 20 to 30 minutes to soften. Squeeze out any excess liquid. Reserve the mushroom soaking liquid, straining it if necessary to remove any grit.

Remove the stems and cut the mushroom tops in half if desired. Slice the bamboo shoots. Peel and finely chop the water chestnuts. Peel the carrots, cut in half, and cut lengthwise into thin strips. Shred the Napa cabbage. String the snow peas and cut in half. Drain the gingko nuts. Crush the ginger.

Combine the reserved mushroom soaking liquid or vegetarian stock with the Chinese rice wine or sherry, dark soy sauce, sugar and sesame oil. Set aside. Heat the wok over medium-high to high heat. Add 2 tablespoons oil to the heated wok. When the oil is hot, add the carrots. Stir-fry for 1 minute, and add the dried mushrooms and lily buds. Stir-fry for 1 minute, and add the water chestnuts, bamboo shoots, snow peas and ginger. Stir in the shredded cabbage and gingko nuts. Add the bean curd sticks. Add the sauce ingredients and bring to a boil. Cover, turn down the heat and let the vegetables simmer for 5 minutes. Taste and add salt or other seasonings as desired. Serve hot.

10. Braised Tofu (By Rhonda Parkinson)

Ingredients

☆ 1 pkg (14oz) House Premium or Organic Tofu Medium or Soft

☆ 4 dried shiitake mushrooms soaked or 5 white mushrooms, canned

☆ 1 large carrot, cooked

☆ 2 green onions

☆ 2 Tbsp vegetable oil

☆ 2 Tbsp frozen snap peas or edamame

☆ 5 slices ginger root

☆ 1 Tbsp soy sauce

☆ 1 tsp sugar, 1/2 cup chicken stock

☆ 1 tsp cornstarch (mix with 1 Tbsp water as paste), 1/2 tsp sesame oil

Method of Preparation

Cut Tofu into 10 square pieces. Cut shiitake mushrooms in half and white mushrooms into three pieces. Slice carrot thinly. Cut green onions into 2-inch pieces. Heat 2 Tbsp oil in wok. Fry Tofu until golden brown. Slide Tofu to the side of wok. Add green onions, shiitake mushrooms, and ginger. Fry until aroma emerges. Add soy sauce, sugar, and stock. Cover and cook over medium heat until the Tofu absorbs the flavour. Add carrots, white mushrooms and snap peas in wok. Thicken with cornstarch paste and splash sesame oil on top. Serve hot.

11. Mushroom Manchurian Dry (By Rhonda Parkinson)

Ingredients

☆ 4 tablespoons Corn Flour

☆ 2 tablespoons Maida Flour

☆ 250 gms (1/2 lb) Fresh Mushrooms (white button mushrooms)

☆ 1/2 teaspoon Garlic Paste

☆ 1/2 teaspoon Ginger Paste

☆ 1/2 teaspoon Soy Sauce

☆ Cooking Oil, Salt, 4 tablespoons Water

For Sauté

☆ 1/2 teaspoon Garlic Paste

☆ 1/2 teaspoon Ginger Paste

☆ 1 Green Chilli, finely chopped

☆ 1 small Onion, finely chopped

☆ 2 tablespoons finely chopped Spring Onion

☆ 2 tablespoons Cooking Oil

☆ $1^{1/2}$ tablespoons Soy Sauce

☆ 1/2 tablespoon Chilli Sauce

☆ 2 tablespoons Tomato Ketchup, Salt

Anyone who likes Mushrooms, will love Mushroom Manchurian to its death. It is rich in Vitamin D and mushroom manchurian dry is one of the best ways to gorge on its goodness while pleasing taste buds with its spicy taste. Additionally, they are rich source of protein and hence it's suitable to be served to kids. It is a great party food and brings a welcome change from routine gobi manchurian.

Method of Preparation

Wash and clean mushrooms to remove any dirt. Dry them with kitchen towel and cut them into medium pieces. Heat oil in a deep kadai (pan). Deep fry them over medium flame until they just started to turn light golden brown. Do not deep fry them long time otherwise they start to release the water and oil may be splatter. Take them out from oil and drain excess oil.

Method for Saute

Heat 2 tablespoons oil in a wide, thin-bottomed, pan (wok) on high flame. Add ginger paste, garlic paste, chopped green chilli and diced onion. Saute on high flame for 1-2 minutes. Add soy sauce, tomato ketchup, chilli sauce, salt, fried mushroom pieces and spring onion; mix them well. Now toss and cook. Keep tossing everything and cook for 1-2 minutes. Delicious dry mushroom Manchurian is ready.

Tips and Variations

After deep frying, do not keep fried mushrooms idle for long time otherwise they may turn soggy because they release the little water after cooking. Chilli sauce and soy sauce forms the basic taste of any Manchurian recipe. You can adjust the quantity of these two sauces to get a desired taste.

Taste
Soft mushrooms covered in crispy outer layer mixed with dry, spicy masala.

12. Mushrooms in Hot Garlic Sauce (By Rhonda Parkinson)
Ingredients
- ☆ 200 gm mushrooms - button, wiped clean
- ☆ 2 Tbsp oil, 1/2 cup onions-chopped fine
- ☆ 3 Tbsp garlic, chopped fine and ground with 2-3 whole red peppers (paprika)
- ☆ Spring onion leaves chopped fine for garnish

Mix together for Sauce
- ☆ 2 Tbsp corn flour - blended with water
- ☆ 1 ¹/² Tbsp salt, 1/4 cup vinegar
- ☆ 1 tsp soya sauce, 1 Tbsp sugar
- ☆ 3/4 cup water

A simple and hassle free recipe. Mushrooms tossed with onions and red peppers, mingled with a garlicky sauce. Serve with a steaming plate of rice.

Method of Preparation
Heat the oil in a wok and add the onions, stir over high heat till a little soft. Add the garlic paste and red peppers and stir-fry till fat separates. Add the mushrooms and stir-fry over high heat, till well mixed. Add a cup of water, bring to a boil and add the sauce mixture and bring to a boil, stirring all the time. Lower the heat and simmer for a minute or so and serve hot garnished with the greens.

12.5. Indian Mushroom Recipe

12.5.1. Soups and Starters

1. Mushroom and Barley Soup (Soups and Salads Recipe) (by Tarla Dalal)
Ingredients
- ☆ 1 tbsp olive oil
- ☆ 3/4 cup chopped onions
- ☆ 2 tsp finely chopped garlic
- ☆ 3/4 cups mushrooms stem, removed and sliced
- ☆ 1/2 tsp freshly ground black pepper powder salt to taste
- ☆ 4 cups vegetable stock
- ☆ 1 cup soaked and half-cooked barley
- ☆ 2 tbsp freshly chopped parsley
- ☆ 1/2 tbsp lemon juice.

Method of Preparation

Soak the barley in enough water for 30 minutes. Drain and keep aside. Boil 3 to 4 cups of water, add the barley and cook on a medium flame for 8 to 10 minutes or till it is 50 per cent cooked. Strain and keep aside. Heat the oil in a deep non-stick pan, add the onions and garlic and sauté on a medium flame for 3 to 4 minutes. Add the mushrooms, pepper and salt and sauté on a medium flame for another 5 to 7 minutes till the mushrooms are tender. Add the vegetable stock, barley and 1 cup of water, mix well and simmer on a slow flame for 10 to 12 minutes or till the barley is completely cooked. Remove from flame, add the parsley and lemon juice and mix well. Serve hot.

2. Pumpkin Mushroom Soup

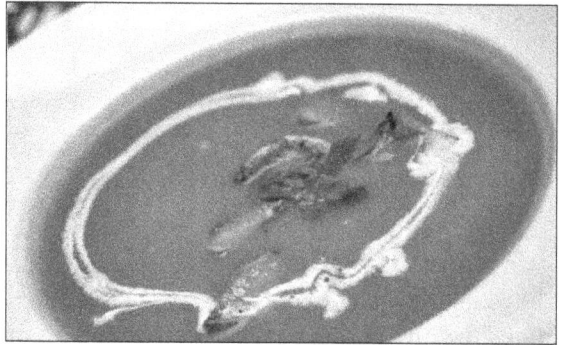

Ingredients

- ☆ 100 g Pumpkin
- ☆ 200 g Mushrooms
- ☆ 80 g Onions (chopped)
- ☆ 10 g Butter
- ☆ 5 g Corn flour
- ☆ 5 g Coriander Powder
- ☆ 1 g Nutmeg
- ☆ 2 cups Stock or water
- ☆ 1 ml Honey
- ☆ 1 cup Milk

Method of Preparation

Sauté mushroom and onions in butter. Add the flour and coriander powder, stirring constantly. Gradually add the stock/water. To this add cooked and mashed pumpkin, salt, pepper, honey and nutmeg. Cook stirring for about 15 minutes add the milk, heat without boiling and serve hot.

3. Sweet and Sour Mushrooms Soup

Ingredients

- ☆ 100g Mushrooms
- ☆ 25 g Cauliflower
- ☆ 10 g French beans
- ☆ 25 g Carrots
- ☆ 25 g Capsicum
- ☆ 15 g Peas
- ☆ 40 g Pineapple
- ☆ 20 g Corn flour
- ☆ 15 g Maida
- ☆ 150 g Tomatoes
- ☆ Oil for frying
- ☆ 2 tsp Salt

Method of Preparation

Clean and cut the vegetables into cubes. Make a batter with corn flour, maida, salt and water. Cook the tomatoes and make a puree. Dip the mushrooms, vegetables and pineapple slices in the batter and deep fry till golden brown. Add the tomato puree and cook the vegetables. Add pine apple juice till the sourness of tomatoes and sweetness of pineapples juice blends.

4. Mushroom Manchurian (by Dassana Amit)

Ingredients

For the Batter

☆ 1 cup all purpose flour/maida

☆ 3 tbsp cornflour,1 tsp ginger garlic paste

☆ 1/2 tsp black pepper powder or freshly crushed black pepper

☆ 1 tsp soya sauce, ¾ cup water,

☆ salt and sugar as required for the manchurian:

☆ 200-250 gms button mushrooms

☆ 3-4 spring onions, finely chopped - reserve the greens for garnish

☆ 2-3 green chillies, finely chopped

☆ 2 tsp finely chopped garlic

☆ 2 tsp finely chopped ginger

☆ 1 tsp finely chopped celery (optional)

☆ 1/2 tsp black pepper powder or crushed black pepper

☆ 1 tbsp soya sauce

☆ 1 medium size green or yellow bell pepper, sliced, diced or chopped, salt as required

Method of Preparation

Preparing the Mushrooms

Rinse, wipe and halve the mushrooms.

Mix everything for the batter. Heat oil for frying. Dip the mushroom in the batter and fry them till golden brown. Keep these aside

Preparing the Sauce

Heat oil. Add the spring onion whites and stir fry them for a minute on a medium flame. Now add the chopped celery, green chilli, ginger, garlic and some of the spring onion greens. Stir fry these also for a minute on a medium flame. Add the black pepper, salt, sugar and soya sauce. Mix well. Add the fried mushroom to this sauce. Stir so that the sauce coats the mushroom well. Serve mushroom manchurian hot garnished with chopped spring onion greens and celery.

Notes

The black pepper, sugar and soya sauce can be adjusted to your taste preferences. so is the quantity of ginger, garlic and green chillies can be increased or decreased with your spice preferences.

5. Capsicum and Mushrooms with Oregano (by Tarla Dalal)

Ingredients

- ☆ 1/2 cup thinly sliced capsicum
- ☆ 1 1/2 cups button mushrooms, cut into halves
- ☆ 1 tbsp butter, 1 tsp chopped garlic
- ☆ salt to taste, 1 tsp Tabasco sauce
- ☆ 1 tsp dried oregano, 1/2 tsp chilli flakes.

For the Garnish

2 tbsp chopped spring onion greens.

The chewy texture of mushrooms and the crispness of capsicum are a good combination, especially when they are sautéed minimally such that they retain their characteristics. Apart from using aromatic oregano, this preparation of Capsicum and Mushrooms with Oregano also stands out in its use of Tabasco sauce, which lends the dish the strong and spicy flavour of special Tabasco capsicums.

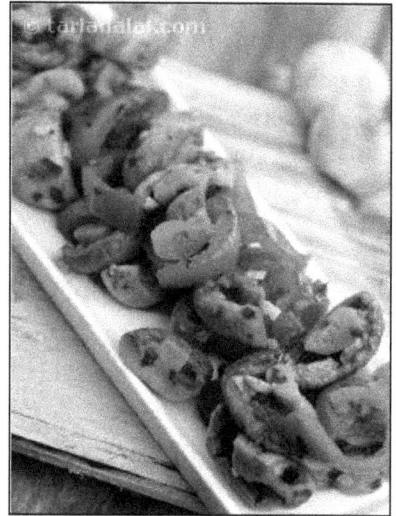

Method of Preparation

Heat the butter in a broad non-stick pan, add the garlic and capsicum and sauté on a medium flame for 2 minutes. Add the mushrooms and salt and cook on a medium flame for 2 to 3 minutes, while stirring occasionally. Add the Tabasco sauce, oregano and chilli flakes, mix well and cook on a medium flame for 1 more minute. Serve immediately garnished with spring onion greens.

6. Hakka Mushrooms (by Tarla Dalal)

Ingredients

- ☆ 2 cups mushrooms cut into halves
- ☆ 4 cloves garlic finely chopped
- ☆ 1 green chilli, finely chopped
- ☆ 2 1/2 tbsp soy sauce, 1 tsp cornflour
- ☆ 1 cup finely chopped spring onions with greens
- ☆ a pinch of chilli powder
- ☆ 1 tsp oil, salt to taste

Method of Preparation

Dissolve the cornflour in 2 tablespoons of

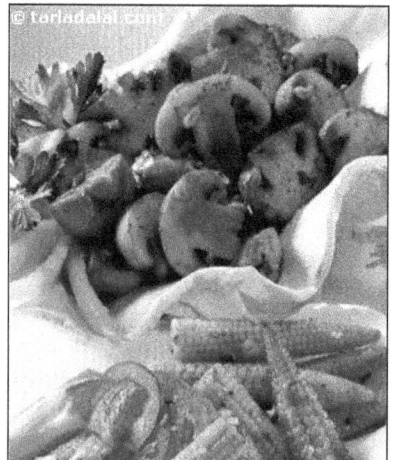

water and the soya sauce and keep aside. Heat the oil in a non-stick pan, add the garlic and green chilli and sauté for 2 minutes. Add the mushrooms and salt and sauté for 3 to 4 minutes. Add the cornflour mixture and sauté for 2 to 3 minutes till the sauce coats the mushrooms. Add the spring onion greens and mix well. Sprinkle the chilli powder and serve immediately.

7. Tandoori Mushrooms (by Tarla Dalal)

Ingredients

- ☆ 300 gms mushrooms cut into halves
- ☆ 1/2 tsp cornflour
- ☆ 1/2 cup low fat milk
- ☆ 1/2 tsp dried fenugreek leaves
- ☆ 1/4 cup low fat curds beaten
- ☆ salt to taste

To be Grinded into a Smooth Paste

- ☆ 4 whole dry Kashmiri red chillies
- ☆ 4 large sized cloves of garlic
- ☆ 25 mm (1") piece of ginger
- ☆ 2 tsp coriander-cumin seeds powder
- ☆ salt to taste

Method of Preparation

Wash the mushrooms thoroughly. Drain and keep aside.Dissolve the cornflour in the milk and keep aside. Heat a non-stick pan on a medium flame and when hot, add the prepared paste and dried fenugreek leaves and cook while stirring continuously for 1 minute. Sprinkle little water if the mixture becomes too dry. Add the mushrooms, cornflour-milk mixture, curds and salt and sauté for 4 to 5 minutes till the mixture coats the mushrooms. Serve hot.

12.5.2.Snacks

1. Mushroom Tikka (by Dassana Amit)

Ingredients

- ☆ 300 gms mushrooms
- ☆ 1 tsp ginger paste
- ☆ 1 tsp garlic paste a pinch of saffron colour
- ☆ 2 tsp bengal gram flour
- ☆ salt to taste, 1 tsp chilli powder

☆ 1/2 tsp garam masala, 1 tsp oil

☆ 2 tbsp curds 2 tsp chaat masala

For the Garnish : onion rings

Method of Preparation

Combine all the ingredients along with mushroom, except the chaat masala, mix well and keep aside to marinate for half an hour. Roast them in tandoor or grill them in a pre-heated griller till they turn golden brown and crisp. Just before serving, sprinkle some chaat masala. Serve hot garnished with onion rings.

2. Mushroom Aloo Chaat (by Gayatri bellare)

Ingredients

☆ 200 gms mushrooms

☆ 200 gms potatoes

☆ 200 gms tomatoes

☆ 200 gms onions

☆ salt to taste

☆ amchur powder

☆ 1/2 tsp garam masala

☆ 1 tsp chaat masala

☆ coriander leaves for garnishing

For Green Chutney

☆ 2 cups coriander leaves

☆ 3 to 4 green chilly

☆ 20 gms ginger

☆ salt to taste

☆ 1 tsp lemon

For Tamarind Chutney

☆ 100 gms tamarind

☆ 200 gms sugar for Garlic Chutney:

☆ 50 gms garlic, salt to taste

Method of Preparation

Cut the mushrooms into halves and soften in hot water for 5 min. Boil the potatoes, peel and cut into small pieces. Chop the tomatoes, onions and green chillies finely. Mix all the ingredients in a bowl, add lemon, amchur powder, salt, garam masala. Add the green chutney, garlic chutney and tamarind chutney,Sprinkle with chaat masala and coriander leaves.

Method for Chutney

Green chutney: mix all the ingredients and grind it into a fine paste.

Tamarind chutney: soak tamarind in water, grind it into fine paste and sieve and add sugar.

Garlic chutney: make a fine paste of garlic, add little water and add a pinch of salt to taste.

3. Mushroom and Capsicum Sandwich (by Vijayau)

Ingredients

☆ 8 bread slices

☆ 100 gms mushrooms

☆ 100 gms capsicum

☆ 1 small sized onion

☆ 1 cup milk

☆ 1 tsp black pepper powder

☆ salt to taste

☆ 1 tsp butter or ghee.

Method of Preparation

Chop the onion in small slices. Cut mushrooms in small size. Also chop capsicum in small slices, Heat the butter/ghee in a frying pan. Add onion to it.Fry for 2 minutes.Add mushroom and again fry it for 5 minutes.Put the lid on the pan for 2 minutes.Then add capsicum to it and fry for 2 minutes. Add salt and blackpepper and mix well. Add the cup of milk. Allow to boil to 2-3 minutes. Then roast the bread slices. Toast it with the mixture of mushroom and capsicum and serve hot with chilly sauce.

4. Mushroom Burger (by Nitika Jain)

Ingredients

☆ 2 no. burger buns

☆ 2 potatoes, 1 onion

☆ 5 mushrooms

☆ 2 tbsp bread crumbs

☆ 2 tsp plain maida flour

☆ 1/2 tsp black pepper powder

☆ 2 to 3 tsp butter

☆ 1 tsp lemon juice

☆ oil and salt as required.

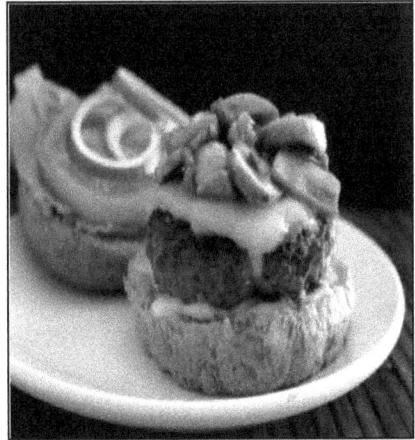

Method of Preparation

Chop the mushroom and keep this aside. Cook and mash the Potatoes nicely and keep this aside. Heat Butter in a pan. Fry Mushroom till it turns into light brown in colour. Add mashed Potatoes, fried mushroom, Onion, 1 plain bread crumbs and salt in a bowl and mix well. Equally divide this mix in to small ball size. Press each ball and flatten them between your palms and make tikkis. Mix All-purpose-flour with Water and apply the paste over the tikkis. Coat tikkis with bread crumbs. Heat a wide pan. Pour 3 to 4 spoons of Oil and spread it over the pan. Keep the stove in the medium heat. Place tikkis and fry till each side turns into brown colour. Fry both the sides. Serve hot with tomato ketchup or place in between toasted burger buns.

5. Mushroom Samosas (by Shahreema)
Ingredients
- ☆ 10 samosa pattis
- ☆ 2 tbsp oil, 1 onion
- ☆ chopped 300 gms diced mushrooms
- ☆ 1 tsp ginger paste
- ☆ 3 tbsp chopped green chillies
- ☆ 2 tsp garam masala,
- ☆ 1/2 tsp cumin seeds powder
- ☆ 2 tbsp chopped coriander
- ☆ 2 tbsp lemon juice, oil for frying
- ☆ salt to taste

Method of Preparation
Heat the oil in a pan and add the onions and saute till they turn golden brown. Add the remaining ingredients, mix well and cook till the mushrooms are soft. Moisten the edges and shape each samosa patti into a cone. Cool and place a portion of the filling in the center and seal the edges like a samosa.Heat the oil in a kadhai and deep fry the samosas till they turn golden brown and crisp. Drain on an absorbent paper. Serve hot.

6. Spinach and Mushroom Pancakes (by sapnil)
Ingredients
- ☆ 2 medium sized bunches spinach
- ☆ 1 medium sized onion
- ☆ 6 to 8 cloves of garlic
- ☆ 8 to 10 medium sized button mushrooms
- ☆ 1 tbsp oil, salt to taste
- ☆ 1/4 tsp white pepper powder
- ☆ 3/4 cup whole wheat flour
- ☆ 3/4 cup low fat milk
- ☆ 1/4 tsp carom seeds

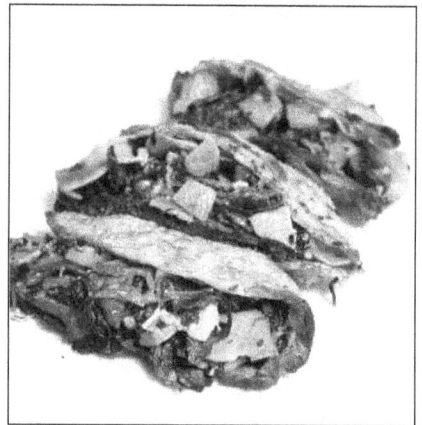

Method of Preparation
Clean and wash spinach leaves.drain and chop roughly. Peel,wash and finely chop onion and garlic. Wash and wipe mashrooms and chop them. Heat oil in a pan add garlic and stir fry briefly. Add onion and mushrooms and cook till onion becomes

soft and translucent cook on high heat. Add spinach,salt and white pepper and cook till moisture dries up. Remove from heat and divide spinach mixture into eight equal portions and keep warm. Mix salt with whole wheat flour and add milk whisk well. Add water as required, to make a smooth batter of pouring consistancy. Strain the batter if there are lumps. Mix in carom seeds and stirr well. Rest the batter for atleast 15 minutes. Heat a nostick pan.grease with oil, if required, Pour half a paddle of batter and spread onto a round shape. Cook for half a minute on medium heat,turn over and cook slightly. Spread the potion of cooked spinach on 3/4 potion of the pancake and then roll it ensuring that the filling does not spill out. Cook rest of the pancakes in similar way.

7. Mushroom and Cheese Dosa (by Rashmirege)

Ingredients

- ☆ 50 gms mushrooms
- ☆ 1 slit green chilli
- ☆ 2 tbsp schezuan sauce
- ☆ salt to taste
- ☆ 3/4 cup grated processed cheese
- ☆ 1/2 tsp black pepper powder
- ☆ 2 cups instant dosa mix

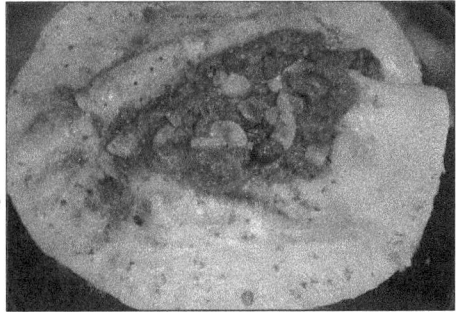

Method of Preparation

Cut the mushrooms and chilli in thin halved slices.In a bowl combine mushroom, chilli, schezuan sauce, cheese, salt and pepper powder together. Heat a non-stick tava (griddle) and sprinkle a little water on it. It should steam immediately. Pour a ladleful of the dosa batter on the tava (griddle) and spread in a circular motion to make a 125 mm. (5") thin dosa. Smear a little oil along the sides, and immediately spread enough mushrooms and cheese mixture (as much as you like) to layer the dosa and cover with a lid until the dosa is done and the cheese has melted. Repeat with the remaining batter to make 5 more dosas. Serve hot with coconut chutney and sambhar.

Handy Tip

Alternately, you could cook the mushrooms first with sauce in a little oil and then spread it on the dosa with cheese.

8. Creamy Mushroom Tartlets (by Tarla Dalal)

Ingredients

For the Tartlets Cases

- ☆ 8 slices of whole wheat bread
- ☆ 1 1/2 tsp melted low fat butter
- ☆ For the Dieter's White Sauce (approx 1 cup) : (¾ cup chopped cauliflower, 2 tsp whole wheat flour, 1/2 cup low fat milk, 1 tsp low fat butter and salt to taste).

For the Creamy Mushroom Sauce
☆ 1 ¹/² cups sliced mushrooms

☆ 3 tbsp chopped onions

☆ 1 tsp finely chopped garlic

☆ 1/2 tsp finely chopped green chillies

☆ 1 cup dieter's white sauce, recipe above

☆ 2 tsp oil and salt and to taste.

Other Ingredients
☆ 8 tsp grated mozzarella cheese

An innovative option for a special tea party. Fill these easy to make whole wheat bread tartlets with delicious low calorie dieter's white sauce that is healthy without compromising on taste. Mushrooms add flavour and texture to this wonderful snack.

Method of Preparation
Remove the crust from the bread slices. Roll each slice with a rolling pin. Press the rolled slices into the cavities of a muffin tray that is lightly greased with butter. Brush with the remaining melted butter and bake in a pre-heated oven at 230°C (460°f) for 8 to 10 minutes or until crisp. Keep aside.

For the Dieter's White Sauce
Boil the cauliflower in 1 cup of water until soft, Blend in a mixer to a smooth puree. Strain and keep aside. Heat the butter in a non-stick pan, add the flour and cook for ½ a minute. Add the milk and cauliflower puree and heat while stirring continuously till the sauce thickens. Add the salt and pepper and mix well. Keep aside.

For the Creamy Mushroom Sauce
Heat the oil in a non-stick pan, add the onions and sauté till they turn translucent. Add the garlic, green chillies and mushrooms and sauté till the mushrooms become soft. Add the dieter's white sauce, salt and pepper and mix well. Simmer for 1 to 2 minutes and keep aside.

How to Proceed
Fill 2 ¹/² tbsp of the creamy mushroom sauce in each tartlet and sprinkle 1 teaspoon cheese on top. Bake in a pre-heated oven at 230°C (460°F) for a few minutes or till the cheese melts. Serve hot.

Handy Tip
You can avoid baking the bread slices in an muffin tray and instead just toast them in an oven or in pop up toaster.

9. Mushroom Balls (by Tarla Dalal)

Ingredients
- ☆ 3 cups button mushrooms
- ☆ oil for deep frying

For the Batter
- ☆ 1 ¹/⁴ cups plain maida flour
- ☆ 1/4 cup cornflour
- ☆ 1 tsp white pepper powder
- ☆ 1 tsp sesame seeds, salt to taste

The speckles of sesame and the hints of white pepper, make this snack a delight to behold and devour

Method for the Batter
Mix together the flour, cornflour, pepper, sesame seeds and salt with ½ cup of water to form a smooth batter. Set aside

Method of Preparation
Wash, drain and fry the mushrooms. Dip the mushrooms in the batter and deep fry till golden brown. Serve hot with sweet and sour dip.

Tips
Make sure the prepared batter is thick enough to coat the mushrooms.

10. Stuffed Mushroom with Paneer (by Tarla Dalal)

Ingredients for the Stuffing
- ☆ 2 tsp butter
- ☆ 1/4 cup finely chopped onions
- ☆ 1 tsp finely chopped garlic
- ☆ 1 tsp finely chopped green chillies
- ☆ 1/2 cup finely chopped mixed vegetables, (sweet corn kernels, tomatoes and capsicum)
- ☆ 1 tsp garam masala
- ☆ 1 cup grated paneer
- ☆ 1/2 cup finely chopped coriander
- ☆ Salt to taste
- ☆ 1/2 tsp lemon juice

Other Ingredients
- ☆ 20 whole large mushrooms de-stemmed and blanched
- ☆ 1 tbsp butter

Mushroom and paneer make an unusual team that will make heads turn. Paneer adds a nice, cheesy feel to the sautéed mushrooms. Ideal as a quick starter for cocktail parties.

Method of Preparation for the Stuffing

Heat the butter in a broad pan, add the onions and garlic and sauté on a medium flame till they turn translucent. Add the green chillies and mixed vegetables, mix well and cook on a medium flame for 2 to 3 minutes, while stirring continuously. Add the garam masala, paneer, coriander and salt, mix well and cook on a slow flame for another 2 minutes, stirring once in between. Add the lemon juice, mix well. Divide the stuffing into 20 equal portions and keep aside.

How to Proceed

Stuff each mushroom with a portion of the stuffing. Keep aside.Grease the butter in a broad non-stick pan and arrange the stuffed mushrooms in it. Cover and cook on a medium flame for 3 to 5 minutes, turning once in between. Serve hot.

Handy Tip

Place the cleaned mushrooms in boiling water for 10 seconds and then refresh them by placing them in cold water. Drain the water and use as required.

12.5.3. Main Course

1. Kadai Mushroom (by Dassana Amit)

Ingredients

☆ 200-250 gms button mushroom

☆ 1 medium or large capsicum, thinly sliced or julienned

☆ 2-3 dry red chilies, roasted or use kashmiri red chilies as they are less hot

☆ 1 tbsp dry roasted coriander seeds

☆ 1/2 tsp garam masala powder

☆ 1 medium size onion, finely chopped

☆ 3 medium size tomatoes, pureed

☆ 1 tsp ginger garlic paste, 1 tsp kasuri methi, roasted and crushed

☆ 2 tbsp oil, salt as required

☆ some ginger juliennes for garnishing, coriander leave for garnishing

Kadai mushroom is an easy and quick dish and yet delicious. Cooked button mushrooms in a semi dry gravy of spiced and tangy tomato sauce along with juliennes of green bell pepper, mushrooms, spices, onions, tomatoes and green bell pepper, their combination never goes wrong.

Method of Preparation

Rinse, wipe and chop the mushrooms. Make the puree of the tomatoes in a blender. Dry roast the dry red chilies and coriander seeds till they get crisp. Don't burn them. The roasting helps to grind them easily. Make a semi fine or fine powder of the red chilies and coriander seeds in a dry grinder or if you have the time and patience in a mortar and pestle. Heat oil. Add the chopped onions and fry them till they transparent. Now add the ginger garlic paste and fry till the raw aroma disappears. Add the tomato puree and the ground red chillies-coriander powder. Saute till the oil leaves the sides of the masala. Add the chopped mushrooms and julienned capsicum and stir. Add ½ to ¾ cup water with salt. Cover and simmer till the mushrooms and capsicum are cooked. Finally add the kasuri methi powder and garam masala powder. Stir and garnish kadai mushroom dish with coriander leaves. Serve kadai mushroom hot with some hot *phulkas, rotis* or *naan*.

Notes

Other varieties of mushrooms can also be used instead of button mushrooms. you can reduce the amount of red chillies to suit your taste buds.

2. Mushroom Peas Curry (by Dassana Amit)

Ingredients

- ☆ 200-250 gms mushrooms
- ☆ 1 cup shelled or frozen peas
- ☆ 1/2 inch ginger and 3 garlic crushed or made into a paste
- ☆ 1 medium size onion finely chopped
- ☆ 1 medium size tomato chopped
- ☆ 1/2 cup grated coconut and 7-8 cashews ground to paste
- ☆ 3/4 tsp red chilli powder
- ☆ 1/2 tsp turmeric powder
- ☆ 1 tsp coriander powder
- ☆ 3/4 tsp garam masala powder
- ☆ 3/4 tsp mustard seeds, 1 tsp cumin
- ☆ 1/4 tsp fenugreek seeds
- ☆ 1 tsp urad dal/split and skinned black gram
- ☆ 1 sprig of curry leaves about 10-12 curry leaves
- ☆ 2 tbsp oil for the curry and 1 tbsp oil for sauting the mushrooms
- ☆ 2 to 2.5 cups stock or water or both, salt as required

Mushroom and peas curry recipe is a rich and delicious recipe of south India. The curry is full on with flavours of coconut and cashews along with the flavours of mushrooms and peas.

Method of Preparation

Boil the peas. strain and keep aside. reserve the stock if using fresh peas. Heat 1 tbsp oil. saute mushrooms in the oil for 5-6 minutes and then keep aside. In another pan, add 2 tbsp oil, add the mustard seeds and let them crackle. Then add the cumin, fenugreek and urad dal. Fry till the oil becomes aromatic and the dal gets browned. don't over brown or burn the dal. Do this tempering on a low or medium flame. Now add chopped onions. fry the onions till light brown. Add the ginger-garlic paste or crushed ginger-garlic. Fry till the raw smell of the ginger-garlic goes away. Now add all the dry spice powders - coriander powder, red chili powder, turmeric powder and garam masala powder. Stir and then add chopped tomatoes. Fry this whole mixture till oil starts to leave the sides. Add the cashew-coconut paste along with curry leaves. Stir for 2-3 minutes. add the stock which we reserved or add water about 2 to 2.5 cups. Let the curry come to a boil. now add the mushrooms and peas. Add salt, stir and let the mushroom peas curry simmer for 5-6 minutes more. Serve mushroom peas curry hot, garnished with some coriander leaves.

3. Curry of Tofu, Mushrooms and Vegetables (by Tarla Dalal)

Ingredients

To be grinded into a Paste

- ☆ 4 to 6 red chillies, 1/4 cup sliced white onions
- ☆ 10 garlic cloves
- ☆ 1/2 tsp cumin seeds
- ☆ 2 tsp coriander seeds
- ☆ 8 black peppercorns
- ☆ 1 tsp grated ginger
- ☆ 3 tbsp coriander leaves
- ☆ 1 lemon, 2 tbsp brown sugar, 1 tsp salt

Other Ingredients

- ☆ 6 big sized fresh mushrooms
- ☆ 1/2 cup boiled green peas
- ☆ 1/2 cup carrot juliennes,boiled
- ☆ 1/2 cup sliced French beans
- ☆ boiled 2 cups coconut milk
- ☆ 1 cup tofu (bean curd/soya paneer) cubes or paneer (cottage cheese) cubes
- ☆ 1/4 cup basil leaves, chopped
- ☆ 1 tbsp soy sauce
- ☆ 1 small sized bundle of lemongrass
- ☆ 1/2 tsp lemon rind, 1tbsp lemon juice
- ☆ 2 tbsp oil, salt to taste.

This Curry of Tofu, Mushrooms and Vegetables is a creamy spicy-sweet curry that speaks of the grandeur of Thai cuisine. While the elaborate masala paste adds spice to the preparation, coconut milk helps balance it with its sweet, milky flavour. Adding lemongrass as a bundle and removing it before serving helps to impart the tangy accents of the herb to the dish while not disturbing the dining experience with tangled shoots of grass!

Method of Preparation

Wash and boil the mushrooms in salted water and drain. Mix with the boiled vegetables. Heat the oil in a pan and add the coconut milk and the paste. Stir well and cool for 4 to 5 minutes till it releases its flavours. Add the mushrooms, vegetables, tofu, basil leaves, soya sauce, tied lemon grass, lemon rind, lemon juice and salt. Simmer for 10 to 15 minutes till the lemon grass releases its juices. Remove the lemon grass bundle. Serve hot with steamed rice.

Tips

When boiling the vegetables, add a pinch baking soda and salt so that they retain their natural colour.

4. Mushroom Curry (by Tarla Dalal)

Ingredients

- ☆ 10 large mushrooms quatered
- ☆ 1 tsp ginger garlic paste
- ☆ 2 tbsp onion paste
- ☆ 1 tsp tomato puree
- ☆ 1/4 tsp chilli powder
- ☆ 3 tbsp curds
- ☆ 1/4 tsp garam masala
- ☆ 1 tbsp oil, salt to taste.

For Garnish : 2 tbsp chopped coriander

Mushroom curry is a completely Indian preparation of mushrooms in a tangy, garam masala flavoured curd base.

Method of Preparation

Heat the oil in a pan, add the ginger-garlic paste and the onion paste and sauté for some time. Add the tomato puree and chilli powder and stir for some time. Add the mushrooms and sauté till they are tender. Finally, add the curds, garam masala and salt and mix well. Garnish with coriander and serve hot.

5. Mushroom Mutter Makhani (Microwave Recipe) (by Tarla Dalal)

Ingredients

For the Paste

- ☆ 1 1/2 cups finely chopped tomatoes
- ☆ 1/2 cup chopped onions
- ☆ 2 large cloves garlic
- ☆ 25 ginger
- ☆ 1 tbsp cashewnuts

Other Ingredients

- ☆ 2 cups mushrooms cut into quarters
- ☆ 1 cup green peas
- ☆ 2 tbsp oil, 1/2 tsp cumin seeds
- ☆ 1/2 cup finely chopped onions
- ☆ 1 tsp chilli powder
- ☆ 1 tsp coriander powder
- ☆ a pinch turmeric powder
- ☆ 2 tbsp fresh cream 1/4 tsp garam masala
- ☆ 1 tsp sugar
- ☆ 1/4 tsp dried fenugreek leaves
- ☆ salt to taste

For the Garnish

1 tbsp chopped coriander

This popular north Indian delight can be made perfectly using the microwave oven, provided you remember a few nuances mentioned here. Make the paste very smooth, ensure that you add the masalas and microwave for five more minutes to harness the full potential of the spices, and take care to discard the water from the marinade to make the vegetables less watery. The outcome will steal your heart and that of your fellow diners.

Method of Preparation for the paste

Combine all the ingredients along with 1/4 cup of water in a microwave safe bowl and microwave on high for 2 minutes.Cool and blend the mixture in a mixer to make a smooth taste. Keep aside.

How to Proceed

Combine the mushrooms with 1/2 cup of water in a microwave safe bowl and microwave on high for 2 minutes. Drain and keep aside.Combine the green peas and 1/2 cup of water in a microwave safe bowl and microwave on high for 3 minutes. Strain and keep aside. Put the oil in a microwave safe bowl and microwave on high for 1 minute. Add the cumin seeds and microwave on high for 1 minute. Add the onions and microwave on high for 2 minutes, stirring once in between after 1 minute. Add the prepared paste, chilli powder, coriander powder and turmeric powder, mix well and microwave on high for 5 minutes. Add the cream, garam masala, sugar, dried fenugreek leaves and ¼ cup of water, mix well and microwave on high for another 1 minute. Add the salt, mushrooms and green peas, mix well and microwave on high for 4 minutes. Serve immediately garnished with coriander.

6. Mushroom Manpasand (by Sammishra)

Ingredients

- ☆ 100 gms mushrooms
- ☆ sliced 1 onion
- ☆ chopped1 tomato
- ☆ 1 tsp chopped ginger
- ☆ 2 tsp chopped garlic
- ☆ 1 tbsp garam masala
- ☆ salt to taste, 1 cup milk
- ☆ 1 tbsp oil

Method of Preparation

Heat the oil in a pan, add the onions and saute till they turn golden brown. Add the garlic, ginger, tomato paste and saute for 2 minutes till oil leaves the side. Add the mushrooms and salt, mix well and sauté for 4 to 5 minutes or till the mushroom is almost half-cooked. Add the milk and mix well and cook in a slow flame till the full mix evaporates and becomes like paneer. Add the garam masala and mix well. Serve hot with paratha.

7. Mushroom Paradise (by Foodie #6528)

Ingredients

- ☆ 1 cup sliced mushrooms
- ☆ 1 cup fresh curds
- ☆ 1/2 tsp freshly ground black pepper powder
- ☆ salt to taste
- ☆ 2 tsp ginger-garlic paste
- ☆ 2 tbsp ghee
- ☆ 1 tbsp oil, 1/2 tsp cumin seeds
- ☆ 2 to 3 cloves
- ☆ 2 to 3 cinnamon sticks
- ☆ 2 to 3 cardamoms, 2 onions sliced
- ☆ 4 chopped green chillies
- ☆ 2 cups soaked long grained rice

☆ 1/2 lemon juice

☆ 3 tbsp chopped coriander

☆ 2 tbsp chopped mint leaves leaves

For the Garnish

A few fried cashewnut

Method of Preparation

Combine the curds, salt, pepper and 1 tsp of garlic-ginger paste in a bowl and whisk well. Add the mushrooms, mix well and keep aside to marinate for 10 minutes. Keep aside. Heat the oil and ghee in a kadhai and add the cumin seeds, cloves, cinnamon and cardamom and saute for few seconds. Add the onions and saute till they turn pink in colour. Add the green chillies, 1 tsp garlic-ginger paste and the mushrooms along with curds and little salt, mix well and cook on high flame till mixture is almost dried. Add rice and lemon juice and 3 ½ cups of water and mix well. Add the coriander and pudina leaves, mix well and cook on slow flame till the rice is cooked. Serve hot, garnish with cashewnuts.

8. Broccoli and Mushroom Bake Dish (Microwave Recipe) (By Tarla Dalal)

Ingredients

For the White Sauce

☆ 1 tbsp butter, 1 ¹ᐟ² tbsp plain maida flour

☆ 3/4 cup milk, Salt and pepper to taste

Other Ingredients

☆ 2 cups broccoli florets, 1 ¹ᐟ² tsp oil

☆ 1 cup sliced mushrooms stalks removed

☆ Salt and freshly ground black pepper powder to taste

☆ 1 tbsp grated mozzarella cheese

The best of continental food on your table for dinner broccoli and mushroom bake is a simple yet tasty preparation of two nutritious foods in white sauce, coated with cheese and baked to perfection. Serve hot, fresh out of the oven, with crispy toasted slices of bread or croutons. Avoid making this ahead of time and reheating as it will ruin the texture.

Method of Preparation

For the White Sauce

Put the butter in a microwave safe bowl and microwave on high for 20 seconds. Add the flour, mix well and microwave on high for 20 seconds. Add the milk, whisk

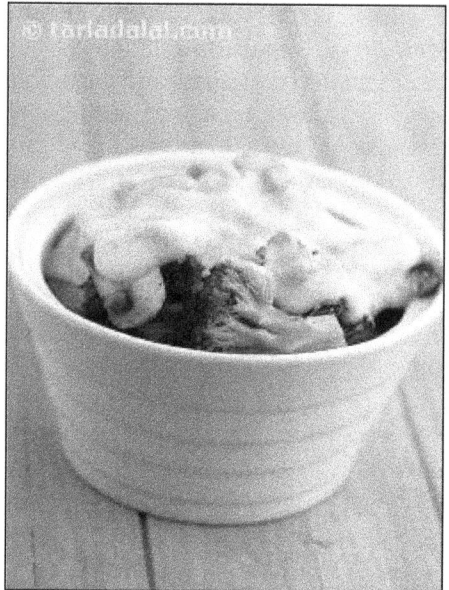

well and microwave on high for 1 minute. Add the salt and pepper, mix well and keep aside.

Method of Preparation

Combine the broccoli with 1 tbsp of water in a microwave safe bowl and microwave on high for 3 minutes, stirring once in between after 1 ½ minutes. Keep aside. Combine 1 tsp of oil and mushrooms in another microwave safe bowl, mix well and microwave on high for 1 minute. Remove from the microwave and add the cooked broccoli, salt and pepper and mix well.

9. Mushroom and Peas in Barbeque Sauce (by Tarla Dalal)

Ingredients

- ✰ 2 cups mushrooms washed
- ✰ 1/2 cup boiled green peas
- ✰ 1/2 barbeque sauce
- ✰ 1 tsp melted butter
- ✰ 1 tbsp cornflour, salt and to taste.

Method of Preparation

Grease the mushrooms with some melted butter, keep them onto a skewer and grill over a charcoal or electric barbeque for 10 to 12 minutes or until the mushrooms are cooked. Mix together the mushrooms, peas and all the ingredients and leave aside for 15 minutes. Wrap in a foil wrapper and grill them over a charcoal or electric barbeque for 10 to 15 minutes. Carefully unwrap the foil wrapper allowing the steam to escape. Serve hot.

Tips

You can also sauté the ingredients in a pan instead of grilling them.

10. Paneer, Mushroom and Capsicum Satay (Kebabs and Tikkis Recipe) (by Tarla Dalal)

Ingredients for the Peanut Sauce

- ✰ 1 tsp oil, 2 tbsp chopped onions
- ✰ 2 cloves
- ✰ 2 tbsp peanut butter
- ✰ 1 tbsp brown sugar
- ✰ 1/2 cup coconut milk
- ✰ 1 stalk lemongrass
- ✰ 1 tsp chilli powder
- ✰ 1/2 tsp grated lemon rind

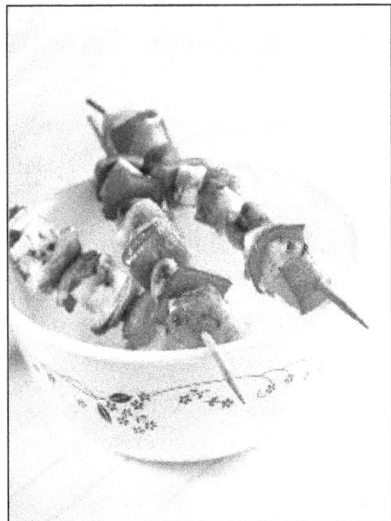

✿ salt to taste, 1 tsp lemon juice

Other Ingredients
✿ 2 cups paneer (cottage cheese) cut into 25 mm. (1") cubes

✿ 1 cup mixed capsicum (red, yellow and green), cut into 25 mm. (1") cubes

✿ 12 large mushrooms

✿ 1 cup onion cubes

✿ oil for cooking

A vegetarian satay of paneer, mushrooms and capsicums marinated in yummy peanut sauce.

Method of Preparation for the Peanut Sauce
Heat the oil in a pan, add the onions and garlic and sauté till the onions are translucent. Add all the remaining ingredients, except the lemon juice and simmer for 10 to 15 minutes. Remove from the flame and discard the lemon grass. Add the lemon juice and mix well. Keep aside to cool.

How to Proceed
Combine ¾ of the peanut sauce with the vegetables and the paneer in a bowl and mix well. Keep aside to marinate for 20 minutes. Thread the paneer, vegetables and onions on 6 wooden skewers. Heat the oil on a non-stick tava (griddle) and cook the skewered paneer and vegetables on all sides fill are golden brown. Serve hot.

Handy Tips
You can use any other vegetables of your choice. You can also grill the satay on a barbecue grill.

11. Tava Mushroom (by Tarla Dalal)
Ingredients
✿ 1 tbsp plain maida flour

✿ 6 cups fresh mushrooms halves

✿ 1 tbsp ghee

✿ 1/2 cup grated onions

✿ 1 tbsp ginger-garlic paste

✿ 1/2 tsp garam masala

✿ 1 tsp chilli powder

✿ 1/2 tsp coriander powder

✿ 1/2 cup fresh tomato puree

✿ 1 tbsp dried fenugreek leaves

✿ salt to taste

✿ 2 tbsp fresh cream

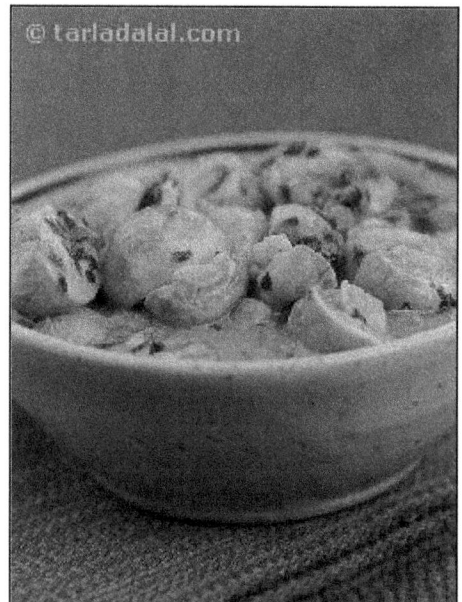

☆ 2 tbsp chopped coriander for the garnish

Cooking on the tava is an age-old custom in the history of indian food. The tava is used not only for making *rotis* and *parathas* but also for preparing delicious dry *subzis*. Tava mushroom is one of favourite recipes that uses basic mughlai masalas such as coriander powder, cumin seeds powder and kasuri methi. Serve this versatile dish as a *subzi* with hot parathas or as a snack with some toast!

Method of Preparation

Sprinkle the plain flour over the mushrooms and keep aside. Heat the ghee in a kadhai, add the onions and sauté till they brown in colour. Add the prepared ginger-garlic paste and sauté again for a minute. Add the garam masala, chilli powder and coriander powder and sauté for another minute. Add the tomato purée and sauté till it thickens. Add the mushrooms, dried fenugreek leaves and salt and cook for a few minutes. Sprinkle a little water while cooking to prevent the masalas from burning. Remove from the flame, add the cream and mix well. Serve hot garnished with coriander.

12. Mushroom Biryani (by Dassana Amit)

Ingredients

☆ 2 cups rice, 4.5 cups water

☆ 200-250 gms mushroom

☆ 2 large tomatoes, 1 large onion

☆ 1 or 2 green chilli, ½ tbsp ginger garlic paste

☆ 1 cup mint + coriander leaves

☆ 1/2 tsp pepper powder, 1 tsp turmeric powder

☆ 2 tsp coriander powder

☆ 1/2 tsp red chili powder

☆ 1/2 or 1 tsp garam masala powder

☆ 1 sprig of curry leaves (optional), salt whole spices:

☆ 1 cinnamon, 1 tsp cumin seeds,

☆ 1 tsp fennel seeds, pinch of mace (optional)

☆ 4-5 cloves, 3-4 small cardamom

☆ 2 bay leaves 1 small star anise

Method of Preparation

Heat oil in a pan. Fry all the whole spices mentioned above till they become fragrant. Now add the onion and fry them till they become transparent. Now add the tomatoes, ginger-garlic paste, curry leaves, green chillies, half of the mint-coriander leaves and all the spice powders. Fry for 2-3 minutes. Now cover the pan and cook the masala. Make sure the masala does not get burnt. Remove after some minutes and stir. When the masala is all one and the tomatoes are all cooked and mushy, add the chopped mushrooms. Saute for 2-3 minutes. Add 4 cups water and stir. Cover the pan with a lid and let the mushrooms cook for some 4-5 minutes. Add rice and mix well. Add salt and the remaining half of the mint-coriander leaves. Cover and cook till the rice is done. While serving mushroom biryani garnish with some mint or coriander leaves.

13. Mushroom Brown Rice (by Tarla Dalal)

Ingredients

For The Brown Rice :

☆ 1 $^{1/2}$ cups brown rice

☆ 3 tbsp chopped lemon grass

☆ 1/4 cup thinly sliced spring onions

☆ 1/4 cup thinly sliced capsicum

☆ 1 tsp finely chopped green chillies

☆ 1 tsp dry red chilli flakes

☆ 1 tsp olive oil or ¼ tsp salt.

For the Sauteed Mushrooms

☆ 4 cups mushrooms chopped into quaters

☆ 1/2 cup sliced spring onions whites

☆ 2 tsp chopped garlic

☆ 1/4 cup thinly sliced yellow capsicum

☆ 1/4 cup thinly sliced green capsicum

☆ 1/4 cup thinly sliced red capsicum

☆ 1/2 tbsp lemon juice

☆ 1 tsp cornflour dissolved in 1 tablespoon water

☆ 2 tbsp chopped coriander

☆ 1 tsp oil, 1 tsp salt.

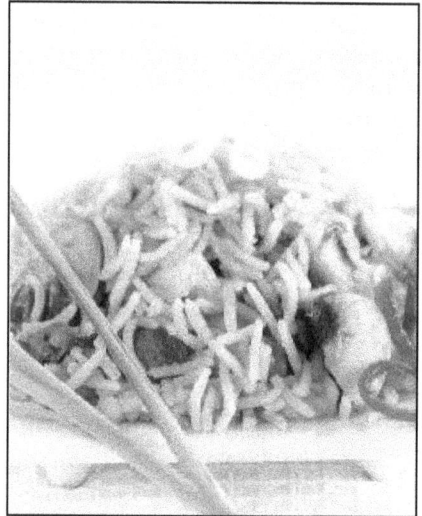

To be Grained into a Paste

3 whole dry kashmiri red chillies, soaked, 2 tsp galangal (thai ginger), 1 tsp sugar, 2 tbsp water.

For fhe Garnish

2 tbsp finely chopped spring onions for the garnish

Method of Preparation

For the Sautéed Mushrooms

Heat the oil in a pan, add the spring onion whites and sauté till they turn translucent. Add the garlic and capsicum and fry for a minute. Add the ground paste, lemon juice and sauté for another minute. Mix well and finally add the mushrooms. Cook till they are soft and all the water has dried. Add the cornflour mixture and coriander and toss gently. Keep aside.

How to Proceed

Just before serving heat the sautéed mushrooms in a large kadhai, add the rice to it and toss gently. Serve hot garnished with spring onion greens.

14. Mushroom Tahini Rice (by Foodie #583160)

Ingredients

- ☆ 200 gms mushrooms
- ☆ 2 cups rice
- ☆ 1 cup sliced onions
- ☆ 2 tsp green chillies
- ☆ chopped 4 tsp ginger
- ☆ chopped 4 tsp garlic
- ☆ chopped 2 tsp cumin seeds
- ☆ 1 stick cinnamon
- ☆ 1 black cardamom
- ☆ 3 peppercorns
- ☆ 2 cloves
- ☆ 2 tbsp sesame seeds
- ☆ 4 tbsp peanuts
- ☆ 1/2 cup curds
- ☆ 1/2 cup lemon juice
- ☆ 4 tbsp vegetable oil

For fhe Garnish

1/2 cup chopped coriander , roasted sesame seeds and peanuts.

Method of Preparation

Wash and soak the rice in 1.2 cups of water. Dry roast sesame seeds and peanuts on non-stick pan and keep aside. After it cools a little, grind and mix with yoghurt to make a paste. Heat oil in a pressure cooker, add cinnamon,cardamom,cloves and peppercorns,sauté and add the cumin seeds. When the seeds crackle add garlic, ginger and green chillies and sauté for 2 to 3 minutes. Add onions and fry till golden in colour.Add salt and sesame - peanut paste, sauté for 1 minute, add mushrooms and the soaked and drained rice, mix well and add 2 cups water, close the lid and cook for one whistle. Allow the steam to escape before opening the lid. Add lemon juice and garnish with chopped coriander and roasted sesame-peanut seeds. Serve hot with pickle.

12.5.4. Mushroom Pickles

Ingredients

- ☆ 1kg Mushrooms small size
- ☆ 6g Cumin seeds
- ☆ 10 g Coriander powder
- ☆ 6 g Onion seeds
- ☆ 30 g Salt and chillis
- ☆ 15 g Ginger
- ☆ 15g Garlic
- ☆ 30 g Onion
- ☆ 8 g fenugreek seed
- ☆ 5g Carom seeds
- ☆ 2 g Cinnamon
- ☆ 4 qty Nutmeg
- ☆ 500 ml Mustard oil
- ☆ Citric acid trace

Method of Preparation

Wash, clean and dry the mushrooms. Sauté in mustard oil till they turn brown and then cool. Now cut onions in to smaller pieces, add ginger, coriander powder, *kalaunji*, salt and spices, *methi, ajwain, dalchini, jaifal* and grind to make a paste. Now mix the ground paste and fried mushrooms. Add vinegar and citric acid. Mix contents well to ensure that the entire contents are submerged with ample amount of mustard oil. Keep in sun for 2 days. Consume as and when required. Always see that the pickle is submerged in oil otherwise it would get spoiled by molds.

12.5.5. Mushroom Chutney

Ingredients

- ☆ 500 g Fresh button mushroom (paste)
- ☆ 2 tsp Vegetable oil
- ☆ 5 finely Chopped green chillies
- ☆ 25 g Ground pepper
- ☆ 2 tsp Garlic paste
- ☆ 1/2 Cup sliced onion
- ☆ 2tsp Salt
- ☆ 2 tsp Lime juice

Method of Preparation

In a large skillet heat oil over medium heat, add mushroom paste and cook mushroom for two minutes. Add green chillies, onion, garlic, black pepper and salt, cook until mushrooms are tender and liquid evaporate.

Glossary

Adnate: Gills joined the stem their entire width.

Adnexed: Gills or tubes narrowly attached to the stipe.

Agaric: Any gill fungus.

AIDS: Acquired Immune Deficiency Syndrome.

Ambient temperature: The natural *i.e.* outside temperature at the time.

Amino acids: A class of organic compounds containing an amino (-NH2) and a carboxyl (COOH) group.

Anaerobic: A microorganism that lives on a process that occurs in the absence of molecular oxygen.

Annulus: The ring around the stem of a mushroom that is formed by the broken veil, remnant of inner veil.

Antibiotic: Chemical compound of an organism that inhibits or kills other microorganisms.

Antidote: A treatment given to counter act the effect of a poison.

Approximate: Of gills which approached but do not touch the stem.

Ascospore: A spore borne in a sac like cell or ascus.

Ascus: The structure containing spores in Ascomycota.

Aseptic: Descriptive of a condition or an object in which unwanted organisms are totally eliminated.

Asexual: Without sex; having the reproductive organs incompletely developed and producing eggs or young by cell-division, budding or parthenogenesis.

Attractants: (1) Materials that attract insects or other animals and cause them to eat or contact poison bait or sprays and consequently causes their death (2) Substance or devices capable of attracting insects or other pests to areas where they can be trapped or killed.

Autoclave: A pressure vessel used to sterilize culture media with steam.

Basidiocarp: Fruiting body of the Basidiomycota which bears basidia.

Basidiospore: A spore from a basidium.

Basidium: A club shaped cell originating as a terminal cell of binucleate hyphae where karyogamy (fusion of nuclei) and meiosis (reduction division) occurs.

Biological efficiency: Biological efficiency is the total yield of mushroom sporocarps in proportion to the dry weight of the substratum at the spawning stage and expressed as percentage.

Button stage: Stage of the mushroom when the cap and the stalk are fully differentiate but are still enclosed by a tissue covering that will later break to expose the gills.

Canning: Method of preservation of foodstuffs including mushrooms. In it suitably prepared foodstuffs (mushrooms) are paced in metal containers that are heated, exhausted and thermetically sealed.

Cap: A part of fruit body which bears spores bearing structures on its under surface.

Carbendazim: Broad spectrum systemic fungicide.

Casing: A layer of material, usually soil or peat moss, placed on the surface of a substrate to stimulate fruit body production.

Chemical control: A control of microorganisms obtained through use of chemical compounds.

Cinnamon: A light brown with a little pinkish.

Cinnamoneous: Cinnamon-coloured.

Clamp connection: A bridge like hyphal connection characteristic of secondary mycelium.

Clitocyboid: More especially of species of Tricholoma having the gills subdecurrent or provided with a decurrent tooth so as to recall Clitocybe.

Competitor mould: Undesirable fungi which compete for the nutrition with mushroom.

Compost: To allow a mixture of substrate, with the help of natural or artificially introduced microorganisms to undergo decomposition to produce a degraded but nutrient rich mixture.

Conceptacle: A hollow case enclosing reproductive bodies.

Concrete: As when the scales adhere to the flesh of the pileus.

Control: (1) Prevention of losses from plant or animal disease, insect pests, weeds etc. by any method (2) To reduce pest population to level below the economic threshold.

Coralloid: Having the form or consistency of coral; especially of the Clavariaceae.

Costa: A longitudinal vein extending along the anterior margin of wing.

Cottony: Covered with a soft cotton like substance.

Cropping: The fruiting stage of the mushroom crop.

Culture medium: A solution or substrate for culturing or growing an organism.

Culture: A growth of one organism for experimental or industrial purpose.

Cup fungi: Ascomycetous fungi with cup shaped ascocarps.

Decurrent: Gills that run down the stem.

Depressed: Having the central portion lower than margin, gills or tubes sunk around the stipe below the general level.

Depressed: Sunk down.

Disease: A condition of structural abnormality or functional failure, which leads to the deleterious effect on mushroom production/economic loss.

Disinfect: To free diseased or infected parts from infection.

Disinfectant: An agent that kills organisms present on the surface.

Dust: A pesticide formulation in dry and finely powdered form used for dusting without further dilution.

Edible fungi: Fruiting bodies of fungi which are fit for eating/human consumption.

Egg: The female gamete or germ cell.

ELISA: Enzyme –linked immune-sorbent assay.

Epigeous: Developing above the ground.

Fistulose: Hollow, like a pipe.

Flabelliform: Fan-shaped.

Flies: A general term for winged insects, but particularly applied to the two winged insects of the order Diptera, the mouth parts of which form a proboscis for piercing and sucking.

Flush: The sudden development of basidiocarps usually occurring in masses and in rhythmic manner.

Form: Shape.

Free: Gills that do not touch the stem.

Fructification: Act of fruiting formation, or fruit formation.

Fruiting body: An organized aggregation of mycelium containing the spore; in Agaricales the mushroom itself composed of the pileus, stipe and gills where the spores are borne.

Fumigants: Volatile chemical applied into confined space.

Fungicide: A substance that kills fungal spores or mycelium.

Fungus: One of the achlorophyllus thallophyte whose somatic structures are usually filamentous and branched. Fungi have cell walls and demonstrable nuclei and reproduce typically by both asexual and sexual means.

Gill: The knife blade like structures on the underside of the pileus of Agaricaceae; lamellae; collectively the hymenophore.

Habit: The general appearance.

Habitat: The place of growth.

Hallucination: Illusion, apparent perception of external object not actually present.

Humidity: Refers to the dampness or amount of moisture in the air.

Hygiene: It covers all the measures which are necessary to give pests and diseases as little chance as possible of developing and spreading.

Hymenium: The layer of basidia or spore bearing cells on the gills.

Hypha: Individual filament of a mycelium.

Incubation: The period after inoculation during which the organism grows.

Infest: To occupy and cause injury to either a plant or to soil, or stored products.

Inoculate: To transfer an organism into a substratum.

Insect pest management: An ecologically based strategy of maintaining insect pest population below the economic injury level by use of any or all control techniques that are economically, ecologically and socially acceptable.

Insects: A class of arthropod with bodies divided into head, thorax and abdomen, the head bearing a pair of feelers or antennae, the thorax, three pairs of legs and wings.

Involute: Rolled inwards.

Lamella: The gills or plates which in Agaricaceae bear the hymenium.

Lateral: Stem attached at one side of the cap.

Lignicolous: Growing on wood.

Lignin: An organic substance forming the woody tissue.

Line: Maintenance of an isolate over time.

Livid: Pale lead-colour.

Longitudinal: Lengthwise.

Marginate: Depressed provided with a narrow, circular, horizontal platform on the upper side.

Marginate: Having a distinctly marked border of the pileus, having a circular ridge on the exterior of the bulb of the stipe, upper angle where the universal veil was attached.

Membranous: Thin and semi-transparent.

Microflora: Vegetation of microorganisms.

Mould: Mycelial micro fungus or a visible growth of such fungus.

Mycelium: A network or mass of hyphae; the vegetative body or structure of the fungus, 'spawns'.

Mycorrhizal: Showing a symbiotic relationship between a mushroom mycelium and the roots of a tree or plant.

Nematodes: Small unsegmented worms, many of which are parasites on plant roots, a member of large group (phylum nematoda) also known as threadworm, roundworm.

Obsolete: Wanting or rudimentary.

Ochreous: Yellow with a tinge of red.

O-Day: Day of stacking.

Oid: A suffix meaning like.

Oxidation: A chemical reaction in which oxygen combines with another substance, or in which hydrogen atoms or electrons are removed from a substance.

Pallid: Somewhat pale.

Paraphysis: Sterile filaments in a hymenium.

Pasteurization: A method of selective destroying unwanted organism usually by heating to a prescribed temperature for a specific period of time.

Patent: Spreading.

Pathogenic: Producing disease or capable of doing so.

Peak heating: Treatments of compost in phase II for killing undesirable micro-organisms.

Petridish: A cylindrical glass or plastic disc with an overlapping cover used for growing microorganisms in a laboratory.

pH: It denotes relative concentration of hydrogen ions in solution; a scale of relative acidity and alkalinity.

Pile or stack: Substrate ingredients formed in a rectangular heap for aerobic fermentation/ composting.

Pileus: The cap structure bearing the hymenium in Agaricaceae.

Pin head: A stage in the development of mushroom at or near the end of differentiation of the cap, but before any enlargement has taken place.

Polymorphic: Having different forms.

PPM: Parts per million. One part per million is equal to 1 milligram in 1 kilogram.

Primordium: An organ in its earliest condition, a basidiocarp initial.

Production rate: Biological efficiency per unit of time.

Pure culture: A culture that contains cells of one kind.

Purification: The separation of organism in pure form.

Radiate: Spreading from a centre.

Receptacle: An axis bearing one or more organs, as the stem in Phalloidaceae; also used for any hymenium – bearing structure.

Relative humidity: The amount of moisture in the air, compared with the maximum amount that the air could hold at the same temperature, expressed as percentage.

Revolute: Rolled back or up of the margin of the pileus.

Rhizomorph: A root like strand of compacted mycelium.

Rudiment: The earliest condition of an organ.

Saprophyte: An organism which lives on dead organic matter.

Simple: Unbroken, unbranched, undivided.

Sinuate: Waved; of gills with a sudden curve before reaching the stem.

Spawn run: Vegetative growth of the mycelium throughout the substrate used to produce mushrooms.

Spawn: A mycelium growing on a substrate used as planting material in mushroom cultivation.

Species: One sort of plant or animal, a division of a genus.

Spent compost: The remaining substrate after the mushrooms have been harvested.

Spore: Structure formed by fungi and corresponding to the seeds of plants, capable off germination and reproduction.

Sporophores: Any spore bearing structure.

Spray: A liquid solution or suspension of a material, such as insecticides, fungicides, etc. made available in the form of fine drops, applied to a surface or body by means of a jet or air or steam forced from the minute opening of a sprayer.

Sterigma: The portion of the basidium bearing the spore.

Sterile: Free from living microorganism.

Stipe: A stalk.

Strain: A group of individuals within a species showing some common observable features or characteristics different from the others. Equivalent to race or variety in plants.

Stuffed: A cusion like body in which the perithecia are immersed in many Pyrenomycetes.

Sub: Prefix, Signifying under, below or partly.

Substrate: The material, usually organic, on which mushroom grows.

Tissue culture: A culture directly obtained from the tissue of the fruiting body of the fungus.

Trama: The tissue between the hymenium in the gills etc.

Tube, tubule: the cylindrical, perforation like hollow that bears the hymenium, in the Boletaceae for purposes of description.

Tubercle: A small wart-like protuberance.

Tuberculate: Having tubercles.

Tunnel/ bulk chamber: A closed insulated chamber in which mushroom compost is pasteurized in bulk with the help of stack.

Variety: A subdivision of a species below the level of subspecies; cultivar.

Volva: The tissue enveloping the young sporophores usually ruptured at the apex, leaving a cup-shaped structure at the base of the stem.

Zone: A girdle.

References

Online Retrieved

"California Poison Action Line: Mushrooms". Retrieved 2008-02-18.

"Death due to Galerina". Seattle Post-Intelligencer. 28 December 1981.

"Delicious or deadly? You pick". Scotsman.com. 02/09/2008.

"Mushroom Poisoning Syndromes". NAMA. 2003. Archived from the original on 2008-03-29. Retrieved 2008-08-13.

"Mushroom Toxins". FDA Bad Bug Book. FDA. 9 January 2008.

"Woman died of mushroom poisoning". BBC News. 2010-03-18.

Ainsworth GC, Sparrow FK, Sussman AS., 1973. A Taxonomic Review with Keys: Basidiomycetes and Lower Fungi. In: Sussman AS, editor. The Fungi, an Advanced Treatise Volume IVB. New York and London: Academic Press.

Arora, David, 1986. Mushrooms Demystified. California, USA: Ten Speed Press. p. 679. ISBN 978-0-89815-169-5.

Bano, Z., and Srivastava, H. C. 1962. Studies on cultivation of Pleurotus sp. on paddy straw. Food Sci. 12:363-365.

Benjamin D.R. and Amatoxin S., 1995. Mushrooms: poisons and panaceas – a handbook for naturalists, mycologists and physicians. New York: WH Freeman and Company. pp. 198–214.

Beug, Michael, 2004. "Mushroom Poisonings Reported in 2004". North American Mycological Association Toxicology Committee. Archived from the original on 2008-07-04. Retrieved 2008-08-04.

Bresinsky A and Besl H., 1990. A Colour Atlas of Poisonous Fungi. Wolfe Publishing. pp. 126–129.

Brozen, Reed and Marcus J H., 2008. "Toxicity, Mushroom – Gyromitra Toxin". emedicine. Medscape. Retrieved 2008-08-04.

Calviño, J., Rafael R., Elena P., Daniel N., Dolores G., Teresa C., Javier M., Victor A., Xose M. L. and Domingo S. G., 1998. "Voluntary Ingestion of Cortinarius Mushrooms Leading to Chronic Interstitial Nephritis". Am J Nephrol 18 (6): 565–569.

Centers for Disease Control (CDC), 1982. "Mushroom Poisoning among Laotian Refugees – 1981". MMWR (USA: CDC) 31 (21): 287–8.

Centres for Disease Control and Prevention (CDC), 1997. "Amanita phalloides mushroom poisoning – Northern California, January 1997". MMWR Morb. Mortal. Wkly. Rep. 46 (22): 489–92.

Chang, Andrew, 2008. "Toxicity, Mushroom – Amatoxin". emedicine. Medscape. Chang, S. T. and P. G. M., 1987. Historical records of early cultivation of Lentinus in China Mush. J. Trop.7:31-37.

Chang, S.T. and P. G. M., 1989. Edible Mushroom and Their cultivation. Boca ratton Florida, U.S.A. CRC Press Inc.

Chodorowski Z., Waldman W. and Sein Anand J., 2002. "Acute poisoning with Tricholoma equestre". Prz. Lek. 59 (4–5): 386–7.

Eschelman, Richard, 2006. "I survived the "Destroying Angel"" (blog). Cornell. Retrieved 2008-08-04.

Evans N, Hamilton, A, Bellu-Villalba, M. J. and Bingham. C. (2012). Irreversible renal damage from accidental mushroom poisoning. BMJ 345 : 5262.

Ferchak, J.D., and Croucher J., 1993. Prospects and Problems in Commercialization of Small Scale Mushroom Production in South and Southeast Asia. In: Chang, S.T., J. A. Buswell, and S. Chiu. (eds). Mushroom Biology and Mushroom Products. Hong Kong: Chinese University Press. pp. 321-323.

Ford, Marsha K., Delaney A., Louis L. and Timothy E., 2001. Clinical Toxicology. USA: WB Saunders. pp. 115.

Gover, DW., 2005. "Fungal Toxins and their Physiological Effects". Retrieved 2008-08-13.

Gussow L., 2000. "The optimal management of mushroom poisoning remains undetermined". West. J. Med. 173 (5): 317–318.

Halpern, John and Andrew R. S., 2005. "Hallucinogenic botanicals of America: A growing need for focused drug education and research". Life Sciences (USA) 78 (5): 519–526.

Hayes, W. A. 1969. Microbiological changes in composting wheat straw/horse manure mixture. Mush. Sci. 7:173-86.

Horowitz. B.Z. and R. G. Hendrickson (2013). Mushroom toxicity treatment and management http://emedicine.meds(ape.com/Article/167398-treatment.

Hruby K, Csomos G, Fuhrmann M and Thaler H.,1983. "Chemotherapy of Amanita phalloides poisoning with intravenous silibinin". Hum Toxicol 2 (2): 183–95.

Ian R. H., 2003. Edible and Poisonous Mushrooms of the World. Timber Press. p. 103.

Kishida, E., Sone, Y. and Misaki, A. 1992. Effect of branch distribution and chemical modifications of antitumor (1/3)- b-D-glucans. Carbohydrate Polymers, 17: 89–95.

Koppel. C(1993). Clinical symptomatology and management of mushroom poisoning. Toxicon.31(12) : 1513-40.

Kumiko, Suzuki; Fujimoto H. and Yamazaki M., 1983. "The toxic prnciples of naematoloma fasciculare." Chemical and pharmaceutical bulletin (Japan) 31 (6): 2176–2178.

Lindgren, Jan., 2003. "Theory for why "edible" mushrooms make some people sick" (newsletter). Spore Prints. Puget Sound Mycological Society.

Mantel W., 1965, Die bewertung von wohlfahrtswirkungen, Allgemeine Forstzeitschrift, 20(33): 506-507

Mantel, E. F.K 1973. Indian J. Mush, 1:15-16.

Marmion, V.J. and Wiedemann, T.E.J., 2002. "The death of Claudius". J R Soc Med 95 (5): 260–1.

Mechem, C., 2007. "Toxicity, Mushroom – Disulfiramlike Toxins". emedicine. Medscape.

Mechem, C and Diane F G., 2008. "Toxicity, Mushroom – Hallucinogens". emedicine. Medscape.

Montanini. S; Sinardi, D; Pratico. C; Sinardi, A. U and G. Trimarchi (1999). Use of acetylcusteine as a life saving antidote in Amanita phalloides (death Cap) poisoning. Case report on "patients. Arzneimittel Forschung. 49(12): 1044-1047.

Padwick G.W., 1941. Mushroom cultivation in India. Indian farming,11:363-366

Pitel, Laura, 2010. "Amphon Tuckey died after eating death cap mushrooms picked at botanic gardens". The Times (London).

Reyes, R.G., and Abella E.A., 1997. Mycelial and Basidiocarp performance of Pleurotus sajor-caju on the Mushroom Spent of *Volvariella volvacea*. Proceedings of Internatrional Seminar on the Development of Agribusiness and its Impact on Agricultural Production in Southeast Asia. Tokyo NODAI Press. pp. 491-497.

Reyes,R.G., Abella E.A., Quimio, T.H., Tayamen, M.J.T and Garcia, B.L., 2003. Philippine Wild Macrofungi with Commercial Potential: Continuing Search and Challenge. Transactions of the National Academy of Science and Technology-Philippines. 25 (1): 78-79.

Reynolds, D.R., 1966. Taxonomic Consideration of a Mushroom under Cultivation in UPCA, Philippines.Philippine Agriculturist 49: 58-763.

Royse, D.J., 1995. Specialty Mushrooms : Cultivation on Synthetic Substrates in the U.S.A. and Japan.Interdisciplinary Science Reviews 20(3): 205-214.

Saller, R., Brignoli R., Melzer, J. and Meier, R., 2008. "An updated systematic review with meta-analysis for the clinical evidence of silymarin". Forsch Komplementmed 15 (1): 9–20.

Sinden J.W., 1946. Synthetic composts for mushroom growing, Bull. Pa. Agric. Exp. Stn. 482: 1-26.

Sinden, J. W., and Hauser E., 1953. The nature of the composting process and its reaction to short composting. Mush. Sci. 2: 723.

Sone, Y., Okuda, R., Wada, N., Kishida, E. and Misaki, A., 1985. Structure and antitumor activities of the polysaccharide isolated from fruiting body and the growing culture of mycelium of *Ganoderma lucidum*. Agricultural and Biological Chemistry 49: 2641-2653.

Stamets, Paul, 2000. Growing gourmet and medicinal mushrooms. Random House, Inc. p. 1.

Tharun, G., 1993. Promotion of Mushroom Production and Bioconvesion of Wastes for Income Generation in Rural Areas. CDG-SEAPO's Biotechnology Training Project. In Chang, S.T., Buswell, J.A., and Chiu, S. (eds). Mushroom Biology and Mushroom Products. Hongkong: Chinese University Press. pp. 307-318.

Thomas, K. H., Ramakrishnan, T. S. and Narasimhan, I. L., 1943. Paddy straw mushroom. Madrs. Agric. J. 31: 57-59.

Veronika P., Elisabeth F., Manuela M., Rohrmoser, Gerhard G., and Meinhard M., 1994. "Partial Purification and Characterization of a Toxic Component of Amanita smithiana". Mycologia (Mycological Society of America) 86 (4): 555–560.

Wasson, Gordon, 1986. Persephone's Quest: Entheogens and the Origins of Religion. privately published. p. 131.

Ancillary Literature

Bulletins

In English

1.   A. J. Roy, M. T. Khan, J. P. kanaujia and T.P.S. Bhandari. Mushroom cultivation.

2.   B.L. Dhar, 1994. Farm Design for White Button Mushroom Cultivation. National centre for Mushroom Research and Training, Chambaghat, Solan- 173213, Himachal Pradesh, 26p.

3.   B. Vijay and Yash Gupta, 1994. Cultivation of White Button Mushroom (*Agaricus bisporus*). National Centre for Mushroom Research and Training, Chambaghat, Solan- 173213, Himachal Pradesh, 73p.

4.   D. S. Sohi, 1974. Mushroom Cultivation. Indian Institute of Horticultural Research, Hessaraghatta Lake Post, Bangalore-560089.

5.   H.S. Sohi, 1986. Diseases and Competitor Mould Associated with Mushroom Culture and Their Control. National Centre for Mushroom Research and Training, Chambaghat, Solan-173213, Himachal Pradesh, 12p.

6.   H. S. Sohi, M.S. Bhandal and K. B. Mehta, 1986. Souvenir on Mushrooms. National Centre for Mushroom Research and Training, Chambaghat, Solan-173213, Himachal Pradesh, 100p.

7.   H. S. Garcha, 1980. Mushroom Growing. Punjab Agricultural University, Ludhiana-1410024, 54p.

8.   Indo-Dutch Mushroom Projects, Directorate of Horticulture, Jeolikote, Distt. Nainital Uttar Pradesh, 12p.

9.   K. B. Mehta, 1990. Mushroom Recipes. National Centre for Mushroom Research and Training, Chambhagat, Solan-173213, Himachal Pradesh.

10. Marimuthu, A. S. Krishinamoorthy, K. Sivaprakasam and R. Jeyarajan, 1989. Oyster Mushroom Cultivation. Tamil Nadu Agricultural University, Coimbatore-641003.

11. R. C. Upadhyay, 1990. Cultivation of Oyster Mushroom. National Centre for Mushroom Research and Training, Chambhagat, Solan-173213, Himachal Pradesh.

12. R. P Singh, 1986. Successful Mushroom Production. G. B. Pant University of Agricultural and Technology. Pantnagar-283145, 8p.

13. R. P. Tiwari, 1977, 1983, 1986. Mushroom Cultivation. Indian Institute of Horticultural Research, Hessarghatta Lake Post, Banglore-560089.

14. S. Saxena and R. D. Rai, 1990. Postharvest Technology of Mushroom. National Centre for Mushroom Research and Training, Chambaghat, Solan-173213, Himachal Pradesh.18p.

15. S. R. Sharma, 1994. Diseases of Mushroom and Their Management. National Centre for Mushroom Research and Training. Chambaghat, Solan-173213, Himachal Pradesh, 43p.

16. S. R. Sharma and Kiran Thakur, 1994. Mushroom Cultivation in India. National Centre for Mushroom Research and Training. Chambaghat, Solan-173213, Himachal Pradesh, 18p.

17. Kaul, 1983. Cultivated Edible Mushroom. Regional Research laboratory (C.S.I.R.) Jammu, 56p.

18. Yash Gupta and S. R. Sharma, 1994. Mushroom Spawn Production. National Centre for Mushroom Research and Training, Chambaghat, Solan-173213, Himachal Pradesh.

Books

1.   Bahl, N.1984. Handbook of Mushroom, Oxford and IBH, New Delhi, 123p.

2.   Chadha, K. L. and Sharma, S. R. 1955. Advance in Horticulture. Volume 13, Mushroom. Malhotra Publishing House, New Delhi, 649p.

3. Garcha, H.S.1984. A Manual of Mushroom Growing. PAU Publication, Ludhiana, 54p.

4. Kannaiyan, S. and Ramasamy, K. 1980. A Handbook of Edible Mushroom. Today and Tomorrow's Printers and Publishers, New Delhi, 89p.

5. Kappor, J. N. 1989. Mushroom Cultivation. ICAR Publication, New Delhi, 104p.

6. Marimuthu, T., Krishnamoorthy, A.S. and Jeyarajan, R. 1991. Glimpses of Mushroom Research in Tamil Nadu and Agricultural University. TNAU Publishers, Coimbatore.

7. Purkayastha, R. P. and Chandra, A. 1985. Manual of Indian Edible Mushrooms. Today and Tomorrow's Printers and Publishers, New Delhi,266p.

8. Quimio, S. T., Chang, S. T. and Royse, D. J. 1995. Technical Guidelines for Mushroom Growing in the Tropics. FAO Plant Production and Plant Protection Paper 106, 155p.

9. Sharma, S. R. and Mehta, K. B. 1991. Bibliography of Mushroom Research of India. NCMRT Publication, Solan, 214p.

10. Sharma, S. R. and Thakur, K 1992. Mushroom Workers in the Directory. NCMRT Publication, Solan, 177p.

11. Singh, H. 1991. Mushrooms – The Art off cultivation. Sterling Publishers Pvt. Ltd., New Delhi, 120p.

12. Tewari, S. C. and Kapoor, P. 1988. Mushroom Cultivation: An Economic Analysis Oxford and IBH, New Delhi, 28 p.

Journals

1. Mushroom research Published by Mushroom Society of India, NRCM, Solan (HP).

2. The Mushroom Journal Published by Mushroom Growers Association, 2 St Pauls Street, Stamford, Lines PE92BE.

3. Mushroom News Published by the American Mushroom Institue, 1, massachusetta Ave., NW Suite 800 Washington, DC 2001. Email: mushroomnews@kennet.net.

4. AMGA Journal, Published by Austalian Mushroom Growers Association Ltd., Locked Bag 3,2 Forbes Street Windsor, NSW 2756. Email: infor@amga.asn.au.

5. Mushroom World, Published by Austalian Mushroom Growers Association, 49, Emma St., Suite 101, Guelph, ON, NIE 6x1.

6. The Spawn Run Published by South African Mushroom Farmer's Association. Editor, at (012) 665-22:2, Email: spawn@global.co.za.

Index

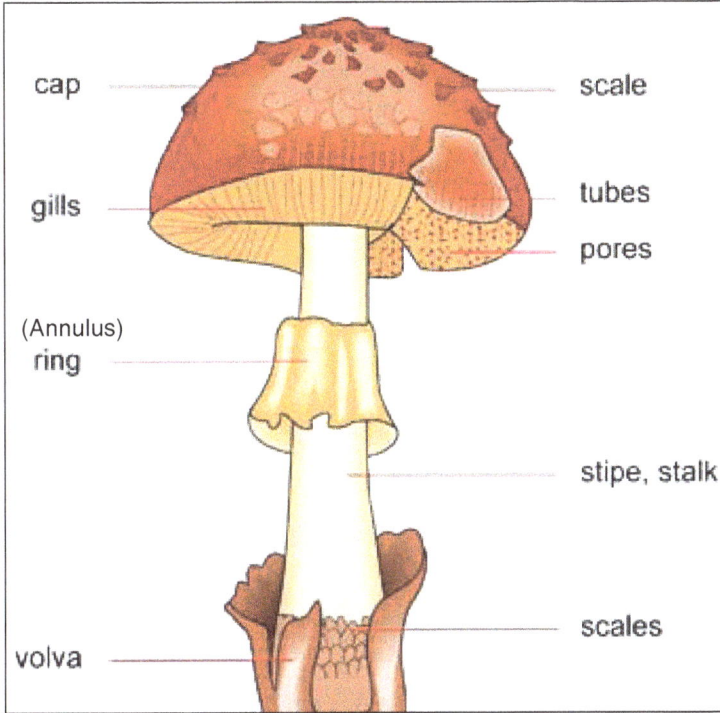

cap — scale

gills — tubes

pores

(Annulus)
ring

stipe, stalk

scales

volva

Cap

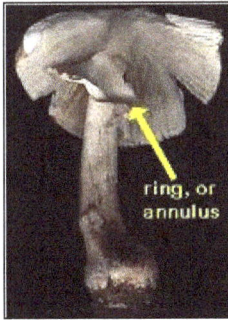

ring, or annulus

Annulus

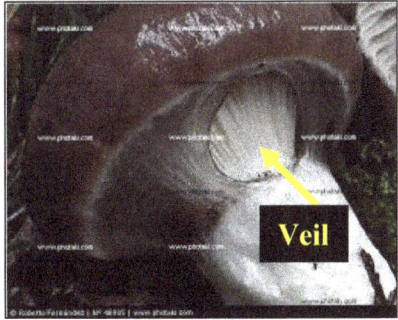

Veil

Veil

Gills

Stipe

Volva

Figure 1: Parts of Mushroom. (p. 10)

Amanita **sp.**   *Agaricus* **sp.**

Volvariella **sp.**   *Marasmius* **sp.**

Figure 3: The Presence or Absence of Annulus and Volva. (p. 13)

Cup fungi (*Peziza* sp.)

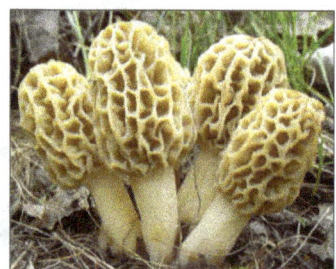

Morels

Figure 4: Epigenous Mushroom. (p. 14)

Truffles

Figure 5: Hypogenous Mushrooms. (p. 14)

Figure 6 (p. 15-16)

I: Humicolous Mushroom

(a) Saprophytic

Lepista nuda

Volvariella **sp.**

Marasmium **sp.**

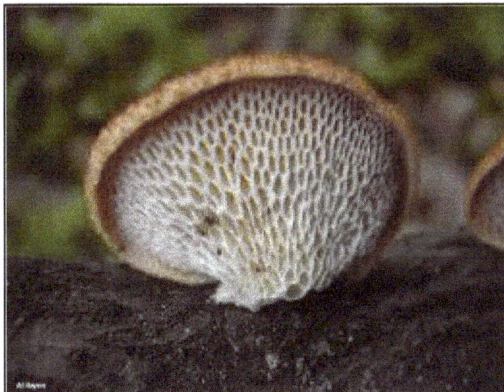

Polyporus tuberaster

(b) Symbiotic

Boletus sp.

Lactarius sp.

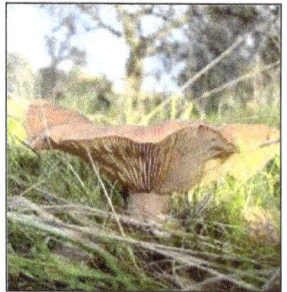

Tricholoma sp.

II. Lignicolous Mushroom

Saprophytic

Agrocybe sp.

Pleurotus sp.

Auricularia sp.

Lentinus sp.

Armillaria

Cyttaria sp.

III. Coprophilous

Agaricus sp.

Coprinus sp.

I. Gill Fungi

II. Pore Fungi

III. Teeth Fungi

IV. Club Fungi

V. Jelly Fungi

Figure 7: Basidiomycetous Mushrooms. (p. 18)

Figure 8: Ascomycetous Mushrooms (p. 19-20)

I. Gilled Mushrooms

Agaricus **spp.**   *Amanita* **spp.**

II. Pore Mushrooms

a) *Fistulina*   **b)** *Trametes*

 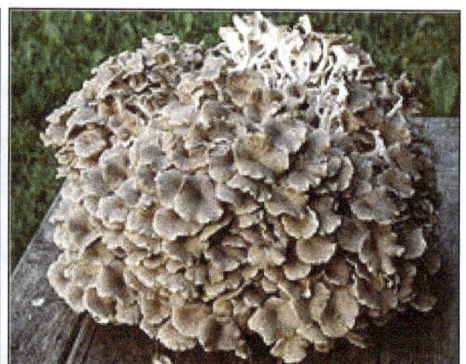

c) *Favolus canadensis*   **d)** *Polyporus*

Contd...

Figure 8–*Contd...*

III. Tooth Fungal Mushroom

a) *Hydnum coralloides*

b) *Hydnum caputursi*

V. Cup Shaped Mushroom

IV. Club Shaped Mushroom

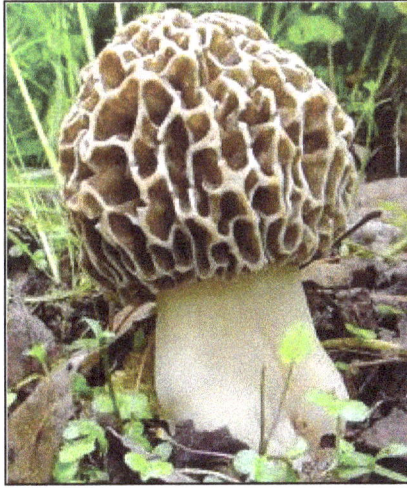

Clavaria **spp.**

Morchella **spp.**

VI. Boletaceous Fungal Mushrooms

a) *Boletinus* **spp.**

b) *Boletus* **spp.**

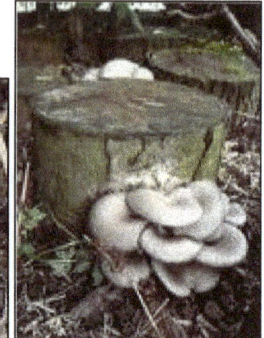

Agaricus **spp.**   *Coprinus* **spp.**   *Pleurotus* **spp.**

Tricholoma **spp.**   *Flammulina* **spp.**   *Calocybe* **spp.**

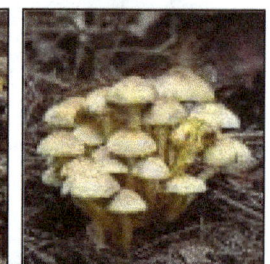

Pholiota **spp.**   *Kuehneromyces* **spp.**   *Hypholoma* **spp.**

Stropharia **spp.**

Figure 9: Mushroom of Subdivision: Basidiomycotina
Order: Agaricales. (p. 21)

Auricularia spp.

Tremella spp.

Figure 10: Mushroom of Order: Auriculariales and Tremella. (p. 22)

Tuber spp.

Figure 11: Mushroom of Subdivision: Ascomycotina, Order: Tuberalles. (p. 23)

Figure 12: Common Edible Mushrooms (Commercially Exploited). (p. 25-26)

Armillaria mellea

Agaricus campestris

Auricularia auricula-judae

Boletus edulis

Cantharellus aurantiacus

Flammulina velutipes

Contd...

Figure 12–*Contd...*

Calocybe indica

Hericium coralloides

Hydnum repandum

Lactarius deceptivus

Lentinula edodes

Marasmius oreades

Morchella spp.

Pleurotus ostreatus

Pluteus cervinus

Tuber spp.

Tremella fuciformis

Volvariella volvacea

Figure 13–*Contd...*

Podaxis pistillaris

Pisolithus arrhizus

Peziza cochleata

Pholiota squarrosa

Psilocybe aeruginosa

Rhizopogon luteolus

Russula aurata

Stropharia aeruginosa

Tricholoma melaleucum

Terfezia **spp.**

Verpa bohemica

Figure 14: Mushrooms Causing Mild to Fatal Reaction and Hallucinogenic Effect. (p. 46-47)

Agaricus xanthodermus

Agaricus pilatianaus

Amanita brunnescens

Amanita muscarisa

Amanita aspera

Amanita phalloides

Boletus purpureus

Boletus rhodoxanthus

Boletus satanas

Boletus satanoides

Clitocybe acromelalgia

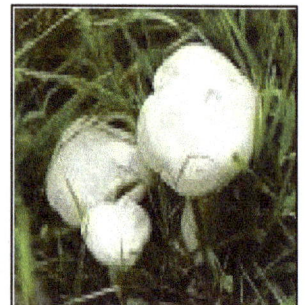

Clitocybe dealbata

Contd...

Figure 14–*Contd...*

Clitocybe rivulosa

Lepota brunneo-incarnata

Paxillus involutus

Psilocybe cubensis

Psilocybe semilanceata

Russula sp.

Tricholoma muscarium

Tricholoma pardinum

Tricholoma sp.

Volvariella media

Volvariella parvula

Figure 17: Commercially Grown White Button Mushroom. (p. 84)

Bag System   **Tray System**

Figure 18: Cultivation Methods of Oyster Mushroom. (p. 86)

Figure 19: Commercially grown Milky Mushroom. (p. 91)

Figure 20: Commercially Grown Paddy Straw Mushroom. (p. 95)

Figure 21: Commercially Grown Shiitake Mushroom. (p. 99)

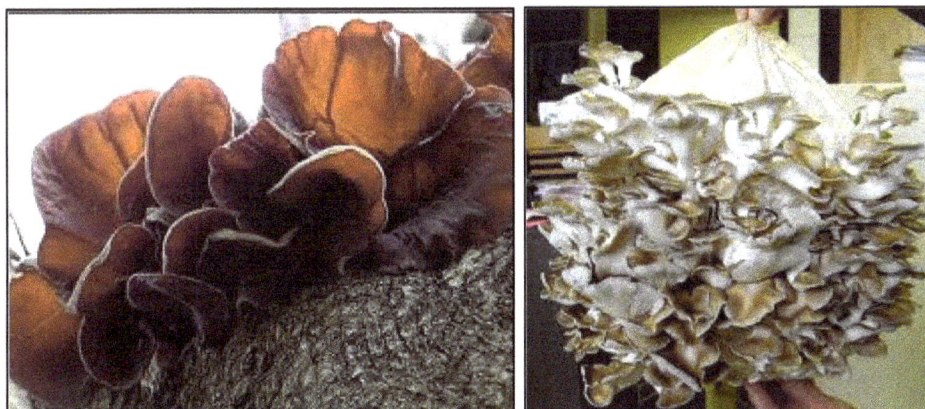

Figure 22: Commercially Grown Black Ear Mushroom. (p. 101)

Figure 23: Commercially Grown Winter Mushroom. (p. 103)

Figure 24: Mushroom Diseases. (p. 109-111)

Dry Bubble

Wet Bubble

Cobweb

Brown Plaster Mould

Contd...

Figure 24–*Contd...*

Green Mould

False Truffle

Olive Green Mould

Ink Caps

Lipstick Mould

Nematodes

Contd...

Figure 24–*Contd...*

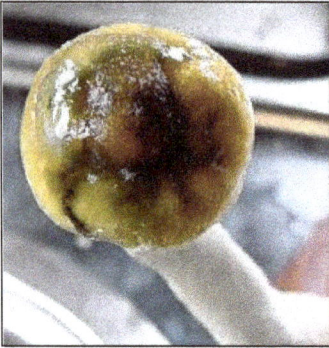

Bacterial Disease

Viral Disease

Physiological Disorders

Rose Comb

Hardening of Gills

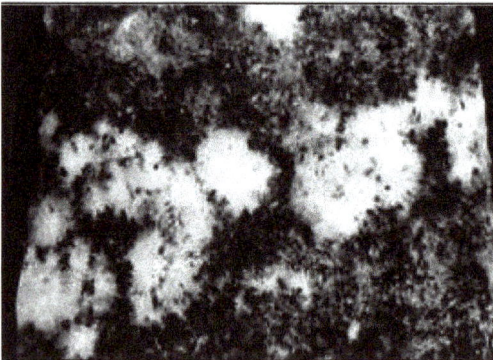

Stroma

Cracking of Caps

Mushroom Recipes

(p. 134)

(p. 154)

(p. 155)

(p. 157)

(p. 158)

(p. 161)   (p. 164)

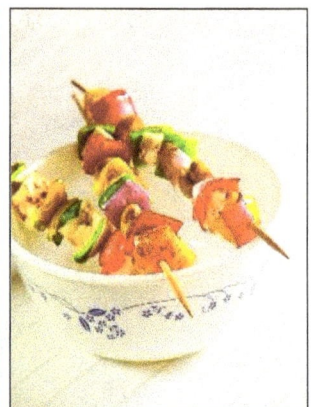

(p. 165)   (p. 169)

www.ingramcontent.com/pod-product-compliance
Lightning Source LLC
Chambersburg PA
CBHW050516190326
41458CB00005B/1554